Applied ethology 2012:
Quality of life in designed environments?

ISAE2012

Proceedings of the 46th Congress of the International Society for Applied Ethology

31 July – 4 August 2012, Vienna, Austria

Quality of life in designed environments?

edited by:

Susanne Waiblinger

Christoph Winckler

Anke Gutmann

ISBN: 978-90-8686-204-7
e-ISBN: 978-90-8686-758-5
DOI: 10.3921/978-90-8686-758-5

First published, 2012

Wageningen Academic Publishers
P.O. Box 220
6700 AE Wageningen
The Netherlands
www.WageningenAcademic.com
copyright@WageningenAcademic.com

Welcome to the 46th Congress of the ISAE

Being the home country of Konrad Lorenz, there is a long tradition for ethological research in Austria. Fostered by societal interest, the field of applied ethology and animal welfare science has gained importance during the last two decades and research divisions at both the University of Veterinary Medicine Vienna and the University of Natural Resources and Life Sciences (BOKU) have successfully been established. The 46th International Congress of the ISAE is now jointly organised by both institutions and we are especially happy to host it for the first time in Austria.

Farm, lab and companion animals as well as wild animals in captivity have to cope with more or less confined conditions. The main topics of the congress therefore centre on the question of 'Quality of Life in Designed Environments'. In this context, we found it relevant to emphasize on positive experiences and emotions. An animal's quality of life does not only depend on the environmental conditions, but also on its experiences during ontogeny, genetic and epigenetic effects constituting the individual disposition to experience negative and positive emotions. Also the social environment - not only within the own species but also regarding the relationships with humans - can contribute to the affective state. Mutilations are still common procedures in animal husbandry and the management of pain associated with it or alternatives to changes in the phenotype receive more interest. Finally, the assessment of welfare and emotions remains a demanding task and innovative methods including cognitive and physiological approaches are being developed.

As for previous congresses the number of submissions exceeded largely the number of slots for oral presentations. In order to draw more attention to the posters, we decided to include sessions with one-minute theatre presentations before the poster viewing sessions. We hope both authors of posters and all congress attendants will enjoy this new feature.

On behalf of the Organising and Scientific Committee we would like to welcome all participants and wish you a fruitful scientific conference with interesting presentations and inspiring discussions.

Susanne Waiblinger & Christoph Winckler

Acknowledgements

ISAE 2012 was jointly organized by

- Institute of Animal Husbandry and Animal Welfare
  Department of Farm Animals and Veterinary Public Health
  University of Veterinary Medicine, Vetmeduni Vienna
- Division of Livestock Sciences
  Department of Sustainable Agricultural Systems
  University of Natural Resources and Life Sciences, BOKU, Vienna

Scientific committee:

Susanne Waiblinger (chair)
Christoph Winckler (vice chair)
Christine Arhant
Johannes Baumgartner
Anke Gutmann
Christine Leeb
Knut Niebuhr
Cornelia Rouha-Mülleder
Claudia Schmied-Wagner
Ines Windschnurer

Organising committee

Susanne Waiblinger (chair)
Christoph Winckler (vice chair)
Anke Gutmann (secretary)
Johannes Baumgartner
Christine Leeb
Daniela Kriebernig
Christina Pfeiffer
Josef Troxler

Ethics committee

Ian Duncan (chair)
Don Lay
Kristin Hagen
Anna Olsson
Marie Jose Hötzel
Alexandra Whittaker
Francois Martin

Technical support

Christian Haberl
Daniela Kottik
Wilhelm Ziegler

Professional conference organisers:

Sophie Wagner at Mondial Congress & Events

Referees

Michael Appleby
Christine Arhant
Johannes Baumgartner
Harry Blokhuis
Xavier Boivin
Oliver Burman
Andrew Butterworth
Elisabetta Canali
Sylvie Cloutier
Mike Cockram
Anne-Marie de Passille
Giuseppe De Rosa
Cathy Dwyer
Sandra Edwards
Hans Erhard
Björn Forkman
Martina Gerken
Anke Gutmann
Lorenz Gygax
Kristin Hagen
Laura Hänninen
Alexandra Harlander-Matauschek
Jörg Hartung
Veronika Heizmann
Edna Hillmann
Gudrun Illmann
Margit Bak Jensen
Linda Keeling
Nina Keil
Ute Knierim
Jan Langbein
Christine Leeb
Lena Lidfors
Xavier Manteca
Jeremy Marchant Forde
Lindsay Matthews
Silvana Mattiello
Marie-Christine Meunier-Salaün
Suzanne Millman
Michela Minero
Lene Munksgaard
Fabio Napolitano
Ruth Newberry
Christine Nicol
Knut Niebuhr
Anna Olsson
Edmond Pajor
Carol Petherick
Bas Rodenburg
Cornelia Rouha-Mülleder
Marko Ruis
Jeffrey Rushen
Vicky Sandilands
Claudia Schmied-Wagner
Barbara Schöning
Lars Schrader
Heike Schulze Westerath
Marek Špinka
Hans Spoolder
Josef Troxler
Cassandra Tucker
Anna Valros
Isabelle Veissier
Antonio Velarde
Kathalijne Visser
Eberhard von Borell
Susanne Waiblinger
Beat Wechsler
Christoph Winckler
Ines Windschnurer

Sponsors

vetmeduni
vienna

University of Natural Resources and Life Sciences, Vienna

BUNDESMINISTERIUM
FÜR GESUNDHEIT

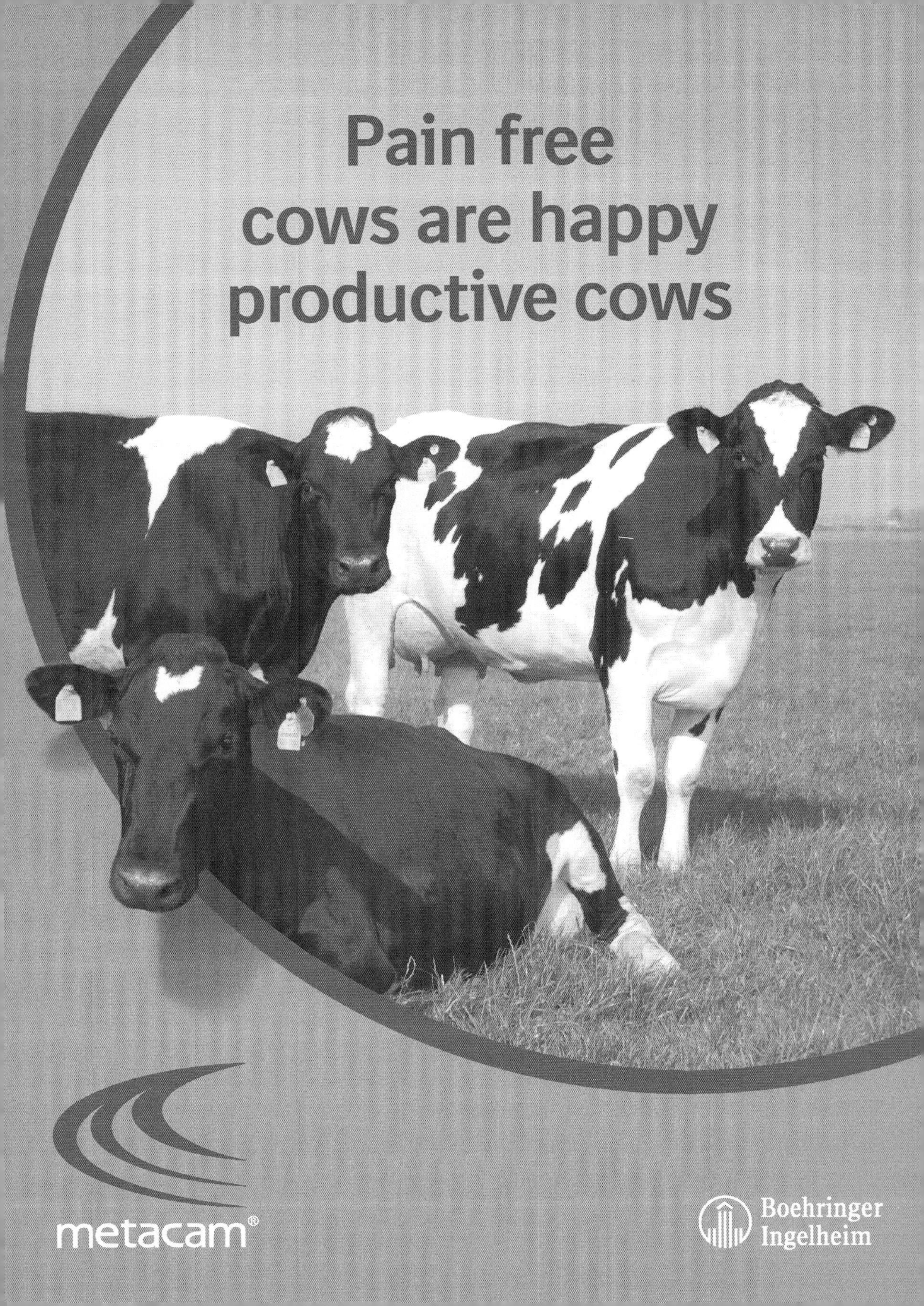
Pain free
cows are happy
productive cows
metacam®
Boehringer
Ingelheim

SERVICE
AnimeD
TIERARZNEIMITTEL

Map of Vienna Underground Lines

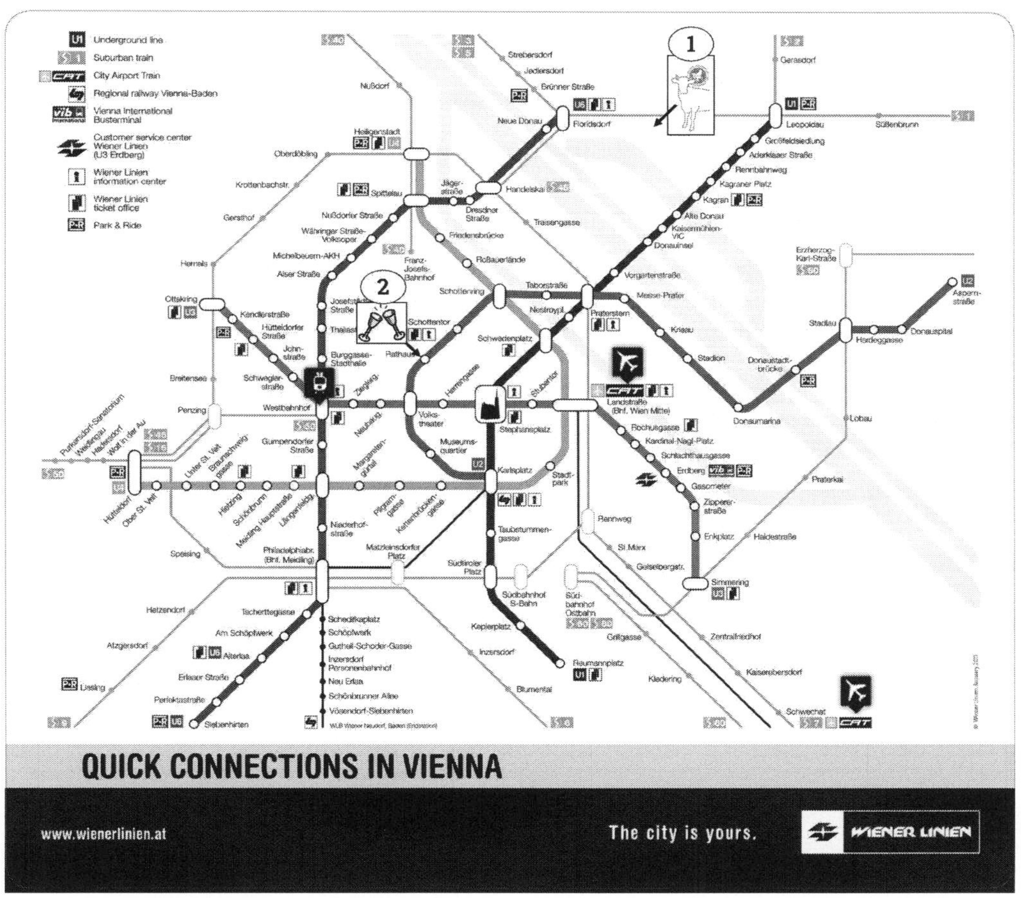

① **Venue** – Vetmeduni Vienna, Veterinärplatz 1, 1210 Wien

② **Welcome Reception**, Vienna City Hall, Rathausplatz 1, 1010 Wien

Map of Venue & Vicinity

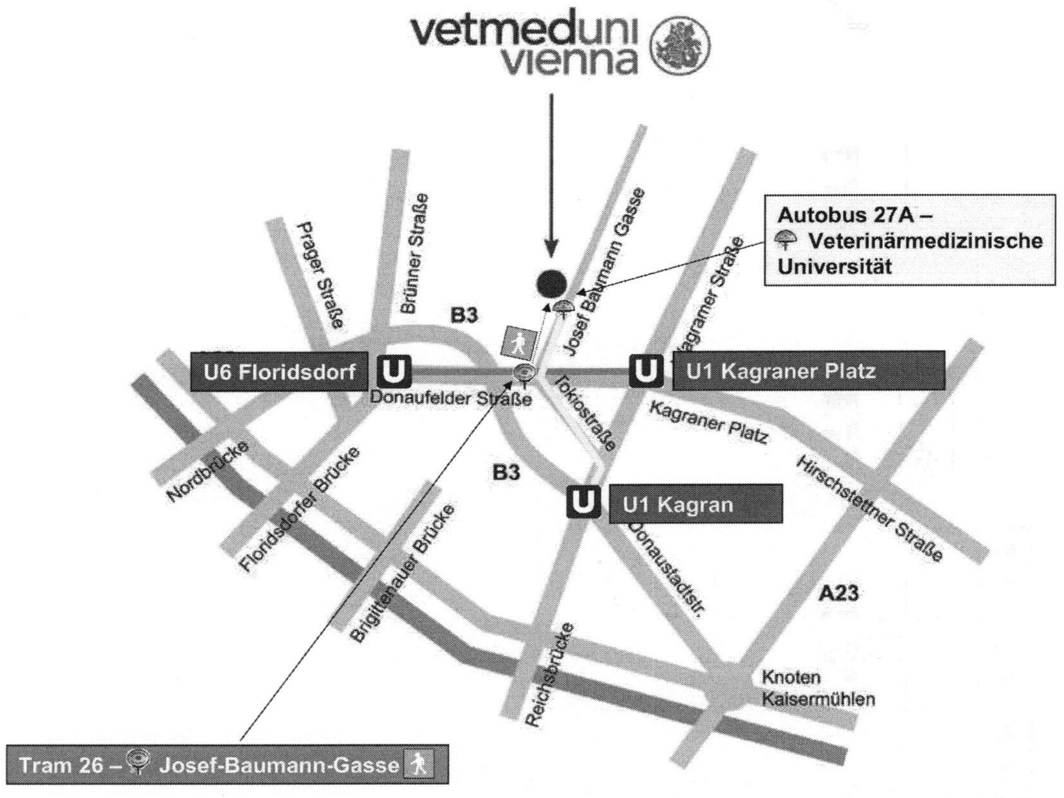

Campus vetmeduni

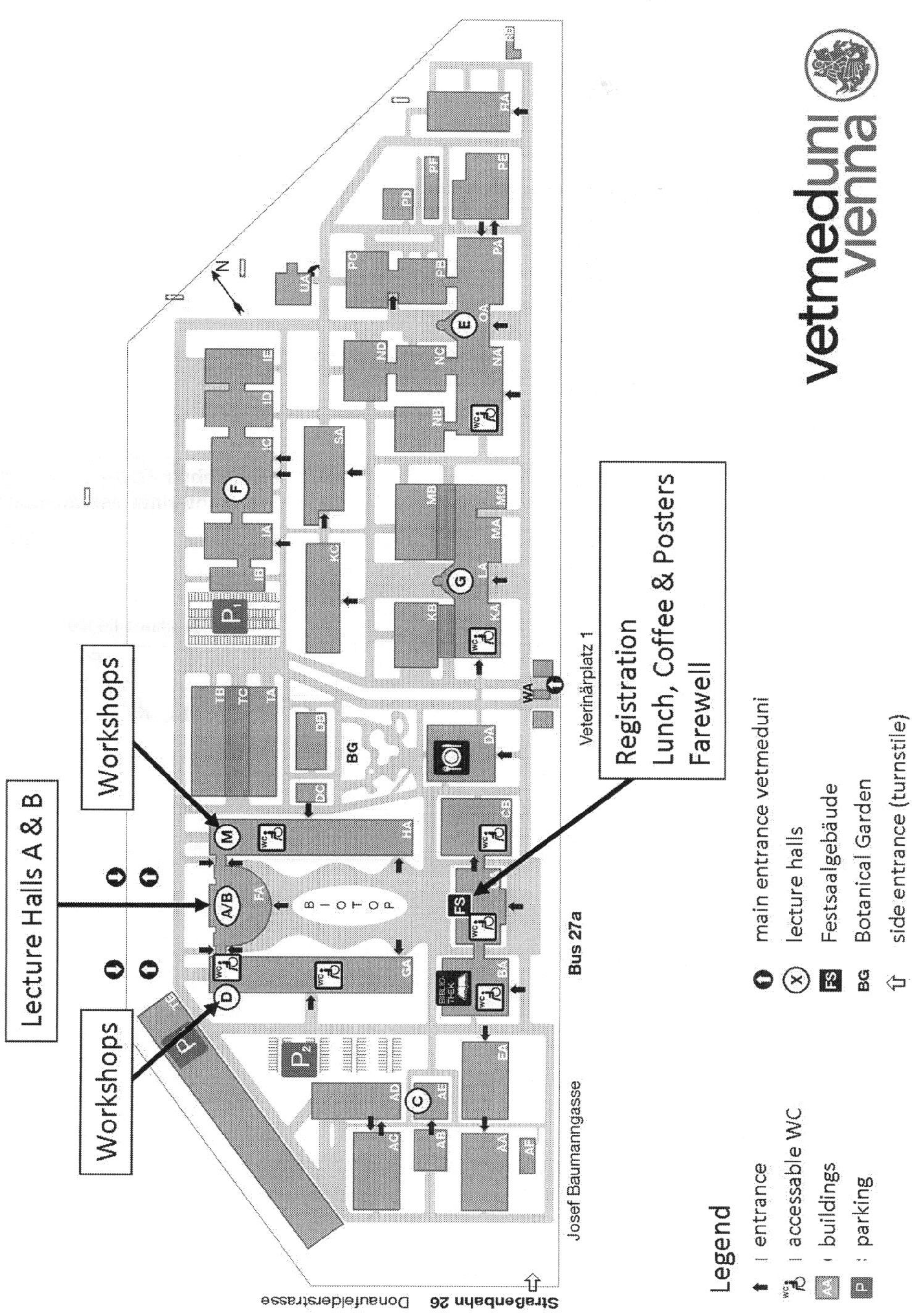

General Information

Venue

The University of Veterinary Medicine Vienna (Vetmeduni Vienna) is located in the 21st district of Vienna. The address is Veterinärplatz 1, 1210 Vienna, Austria (no. 1 on the map).

Official Language

Official language of the meeting is English.

Registration and Information Desk (Festsaalgebäude, ground floor)

Opening hours:

Tuesday, July 31	15:00 – 18:00
Wednesday, August 1	07:30 – 19:00
Thursday, August 2	08:00 – 13:30
Friday, August 3	08:00 – 16:30
Saturday, August 4	08:30 – 17:30

Email: isae2012@mondial-congress.com
Phone: 0043-676-845880701

Name Badges

Your name badge is your admission to the venues, the scientific sessions, poster sessions and to the lunch and coffee breaks. It should be worn at all times at the conference venue and at social events.

Poster and Exhibition Area

The poster and exhibition area will be in the Festsaalgebäude at Vetmeduni campus. We encourage you to visit the exhibition.

Internet Access

Wireless internet access is available free of charge using eduroam (www.eduroam.org).
UserName: isae2012
Password: sAmTLcxa

Coffee Breaks and Lunches

Coffee and refreshments and lunch will be served at the Festsaalgebäude. Light refreshments will also be available in front of the lecture halls.

Welcome Reception: Tuesday, July 31, 19:00 – 21:00

The welcome reception will take place at Vienna City Hall (no. 2 on the map). A buffet and drinks will be served. Vienna City Hall is located at the Wiener Ringstrasse, a circular road surrounding the City Center of Vienna and one of its main sights. The easiest way to go there is using public transport (tram line 1 & D to station Rathausplatz/Burgtheater, tram line 2 to station Stadiongasse/Parlament, or underground line U2 to station Rathaus). You'll receive your entrance ticket with your registration documents or directly at Vienna City Hall.

Excursions: Thursday, August 2, 13:30 – 19:00

(Please note that pre-registration is required.)
All excursion buses will depart at the main entrance of Vetmeduni Vienna at 13:30. A Lunchbox is included.

Conference Dinner: Friday, August 3, 19:00 – 23:00

(Please note that pre-registration is required.)
The conference dinner will be held at Heuriger Wolff, a traditional Eastern-Austrian wine-tavern, where wine-growers serve the most recent year's wine.

Farewell Party: Saturday, August 4, 18:30 – open end

The farewell party will take place in the Festsaalgebäude of Vetmeduni Vienna. A cold buffet and drinks will be served.

Banking service, currency

Euro (€) is the official currency in Austria. Exchange offices are available at the airport or at Stephansplatz in the City Center; the exchange offices are open approximately 6:00 – 23:00. There are plenty of cash dispensers (ATMs) in Vienna including one at Vetmeduni campus.

Shopping in Vienna

Most supermarkets and large stores in Vienna are open 8:00–19:00 on weekdays, and 9:00–17:00 on Saturdays. Smaller, more specific and retail shops have quite variable (and special summer-) opening hours and often have midday-break. On Sundays all stores are closed except shops at Westbahnhof, Franz-Josefs-Bahnhof and Praterstern.

Transport from and to Vienna International Airport

City Airport Train (CAT)

The CAT takes 16 minutes nonstop to get from central Vienna to the airport and vice versa and operates daily 5:38 – 23:35 every 30 minutes. The City Air Terminal is just 10 minutes from St. Stephan's Cathedral at Wien Mitte/Landstrasse station by public transport. Price (single): € 11.00. Ticket machines are on the platforms at the airport and at Wien Mitte/Landstrasse. See http://www.cityairporttrain.com/ for details and online tickets.

S-Bahn (suburban railway)

The Schnellbahn (S-Bahn) is a low-priced way of getting from Vienna to the airport and back. It operates daily 4:31 – 23:46 (Wien Mitte/Landstrasse → Airport) / 4:54 – 0:18 (Airport → Wien Mitte/Landstrasse), approximately every 30 minutes. Price: from €4.00 (including travel on Vienna public transport). Ticket machines are on the platforms at the airport and at Wien Mitte/Landstrasse. See http://www.oebb.at/en/ for details and online tickets.

Regional buses

Vienna Airport Lines (Postbus) connects Vienna Airport with the main Vienna transport hubs. The three Vienna Airport Lines routes serve all of Vienna's underground railway lines and also "Wien Westbahnhof" and "Wien Meidling" railway stations.
To get directly from airport to venue the most convenient connection is to use the Vienna Airport Lines regional bus line 1183 (to Kaisermühlen-Kagran) to the terminal station "Wien Donauzentrum (Hotel)" and change to bus line 27A to bus stop Veterinärmedizinische Universität. Prices (single): from € 8.00. See http://www.postbus.at/en/Airportbus/index.jsp for details.

Taxi

Exit the airport in direction to K3, next to the arrival hall (Tel.: +43-1-7007-35910). See http://www.ats-vie.com/wordpress/ for details.

Emergency calls

112 Euro Emergency call (in Austria: call-through to police)
Austria – country code +43
122 firefighters
133 police
144 rescue

Local Conference Secretariat

Mondial Congress & Events
Operngasse 20b
A-1040 Vienna, Austria
T +43-1-58804-0
F +43-1-58804-185
E isae2012@mondial-congress.com

Programme at a Glance

Tues July 31		
15:00-18:00	Registration & installation of posters	
19:00-21:00	Welcome reception at Vienna City Hall	
Wed August 1	**Lecture Hall A**	**Lecture Hall B**
7:30	Registration & installation of posters	
9:00- 9:20	Opening	
9:20- 9:50	World Organisation for Animal Health (OIE)	
9:50-10:50	Wood Gush Memorial Lecture (Sachser, Norbert)	
10:50-11:20	Coffee	
11:20-11:55	Plenary 01: **Behavioural development** (Jensen, Per)	
12:00-13:00	Session 01: *Behavioural development*	Session 02: *Temperament – Individual differences*
13:00-14:20	Lunch	
14:20-14:55	Plenary 02: **Behavioural development** (Rodenburg, Bas)	
15:00-16:15	Session 01 continued	Session 02 continued
16:15-16:45	Coffee & poster	
16:45-19:15	Parallel Workshops	
19:15-21:30	Drinks & snacks at posters	
Thurs August 2	**Lecture Hall A**	**Lecture Hall B**
8:30- 9:10	Plenary 03: **Intra- and interspecies relationships** (Nowak, Raymond)	
9:15-10:45	Session 03: *Intra- and interspecies relationships*	Session 04: *Behavioural methods*
10:45-11:30	Coffee & posters	
11:30-13:15	Session 03 continued	Session 05: *Environment, behaviour and welfare I*
13:15-19:00	Excursions	
Fri August 3	**Lecture Hall A**	**Lecture Hall B**
8:30- 9:10	Plenary 04: **Positive emotions** (Held, Suzanne & Špinka, Marek)	
9:15-10:30	Session 06: *Positive emotions*	Session 03 continued
10:30-11:15	Coffee & posters	
11:15-12:45	Session 06 continued	Session 03 continued
12:45-14:15	Lunch	
14:15-14:55	Plenary 05: **Cognition & emotion** (Huber, Ludwig)	
15:00-16:00	Session 07: *Cognition and emotion*	Session 08: *Mutilations*
16:15-17:45	AGM – Annual General Meeting of ISAE	
19:00-23:00	Conference Dinner at 'Heuriger Wolff'	
Sat August 4	**Lecture Hall A**	**Lecture Hall B**
9:00- 9:40	Plenary 06: **Behavioural indicators of welfare** (Martins, Catarina)	
9:45-11:00	Session 09: *Behavioural indicators of welfare*	Session 10: *Environment, behaviour and welfare II*
11:00-11:30	Coffee & posters	
11:30-12:30	Session 09 continued	Session 10 continued
12:30-13:30	Lunch	
13:30-14:10	Plenary 07: **Environment, behaviour and welfare** (Ostindjer, Marije)	
14:15-15:00	Session 09 continued	Session 10 continued
15:00-15:30	Coffee	
15:30-16:45	Session 11: *Free papers*	Session 10 continued
16:45-17:15	Closing	
18:30	Farewell Party at Vetmeduni	

Scientific Programme

July 31, 2012 – Tuesday

15:00–18:00	Registration open & installation of posters
19:00–21:00	Welcome reception at Vienna City Hall

August 1, 2012 – Wednesday

07:30	Registration open & installation of posters	
09:00–09:20	Opening	
09:20–09:50	*Vallat, Bernard* World Organisation for Animal Health (OIE)	
09:50–10:50	**Wood Gush Memorial Lecture** (*Lecture Hall A*)	
	Sachser, Norbert The social modulation of behaviour during development	
10:50–11:20	**Coffee**	
11:20–11:55	**Plenary 01:** **Behavioural development: genes, epigenetics and ontogeny** (*Lecture Hall A*)	
	Jensen, Per Behaviour epigenetics – the connection between environment, stress and welfare	
12:00–13:00	**Parallel Sessions**	
	Lecture Hall A	*Lecture Hall B*
	Session 01: Behavioural development: genes, epigenetics and ontogeny	***Session 02:*** Temperament – Individual differences
12:00	*Dwyer, Cathy M.* Prenatal undernutrition affects spatial memory in male lambs in a genotype-dependent manner	*Graunke, Katharina L.* Temperament dependent reactivity of calves in challenging test situations: a multivariate and correlational analysis of behavioural and physiological measures
12:15	*Matthews, Lindsay R.* Association between ewe undernutrition around conception and laterality and emotional responsiveness in the offspring	*Castanheira, Maria F.* Linking cortisol responsiveness and aggressive behaviour in gilthead seabream Sparus aurata: evidence of divergent coping styles
12:30	*Ringgenberg, Nadine* Impacts of prenatal stress and environmental enrichment during lactation on the behaviour and performances of piglets after weaning	*Cussen, Victoria* Perseverative parrots: individual differences and changes in perseveration in orange-winged Amazon parrots (*Amazona amazonica*)
12:45	***Poster Presentation 01:*** Behavioural development / Free Nos. 1-10, 108-109	***Poster Presentation 02:*** Temperament – Individual differences / Environment, behaviour and welfare Nos. 11-16, 29-34
13:00–14:20	**Lunch**	

14:20–14:55	**Plenary 02:** **Behavioural development: genes, epigenetics and ontogeny** (*Lecture Hall A*)	
	Rodenburg, Bas The role of genes, epigenetics and ontogeny in behavioural development	
15:00–16:15	**Parallel Sessions**	
	Session 01 continued: Behavioural development: genes, epigenetics and ontogeny	***Session 02 continued:*** Temperament – Individual differences
15:00	*Ericsson, Maria* Domestication effects on stress recovery in chickens (*Gallus gallus*)	*Ison, Sarah H.* Relationships between behaviour during temperament tests in prepubertal gilts with farrowing behaviour in two environments: a conventional crate or loose-housed (PigSAFE) pen
15:15	*Wagner, Kathrin* Long term effects of mother rearing: challenge responses of primiparous dairy cows	*Melotti, Luca* Response strategy of pigs in a delay discounting task is related to homovanillic acid excretion in urine but not to coping styles or aggression
15:30	*Telkänranta, Helena* The effect of post-natal environmental enrichment on tail biting in growing pigs	*Dimitrov, Ivan* Assessing emotional processes in dairy sheep of different temperaments: Behavioural, cortisol and immune responses in dairy sheep experiencing positive and negative emotions on farm
15:45	*Hopkins, Jessica E.* The influence of neonatal environment on piglet play behaviour and post-weaning social and cognitive development	***Poster Presentation 04:*** Environment, behaviour and welfare Nos. 35-57
16:00	***Poster Presentation 03:*** Behavioural indicators of welfare Nos. 17-28	
16:15–16:45	**Coffee & posters**	
16:45–19:15	**Parallel Workshops**	
16:45–19:15	***Workshop 1:*** *Tucker, Cassandra and Yngvesson, Jenny* Evaluation of student learning	
16:45–19:15	***Workshop 2:*** *Špinka, Marek et al.* Farm animal welfare research and education in an enlarged Europe and beyond	
19:15–21:30	**Drinks & snacks at posters**	

August 2, 2012 – Thursday

08:30–09:10	**Plenary 03:** **Intra- and interspecies relationships** (*Lecture Hall A*)	
	Nowak, Raymond Social attachment in sheep: similarities and differences between intra- and interspecies relationships	
09:15–10:45	**Parallel Sessions**	
	Lecture Hall A	*Lecture Hall B*
	Session 03: Intra- and interspecies relationships	***Session 04:*** Behavioural methods
09:15	*De Oliveira, Daiana* Early tactile stimulation of piglets produces different subsequent behaviour responses correlated to body development	*Rushen, Jeffrey* Measures of acceleration can estimate the effects of dehorning and weaning on the duration of locomotor play in calves
09:30	*Coulon, Marjorie* How do lambs perceive regular human stroking? Behaviour, cortisol and heart rate variability analyses	*Buijs, Stephanie* Open-field behaviour in young rabbits – How much do the lines tell us?
09:45	Laffitte, Béatrice Tying dairy calves for improving human-cows relationship seems not justified	*D'Eath, Richard B.* Three kinds of agreement: bias, association and exact matching
10:00	*Mersmann, Dorit* Influences on avoidance and approach behaviour of dairy goats towards an unfamiliar person	*Nielsen, Per P.* Automatic registration of grazing behaviour in dairy cows
10:15	*Rault, Jean-Loup* A positive mindset in the face of stress	*Mackay, Jill R. D.* Novel object tests inform on home pen activity patterns in dairy cattle
10:30	***Poster Presentation 05:*** Intra- and interspecies relationships Nos. 69-80	***Poster Presentation 06:*** Environment, behaviour and welfare Nos. 58-68
10:45-11:30	**Coffee & posters**	

11:30–13:15	**Parallel Sessions**	
	Session 03 continued: Intra- and interspecies relationships	***Session 05:*** Environment, behaviour and welfare I
11:30	*Cloutier, Sylvie* Tickling makes handling a more pleasurable experience for laboratory rats	*Harlander-Matauschek, Alexandra* Do chopped feathers in pelleted diets of chicks reduce the appetite for feathers from conspecifics?
11:45	*Gouveia, Kelly* Enhancing mouse welfare through non-aversive handling	*Riber, Anja B.* Can heterogeneity of nest boxes reduce gregarious nesting in laying hens?
12:00	*Nielsen, Birte L.* In search of oestrus odours: do certain smells elicit penile erections in sexually naïve rats?	*Gilani, Anne-Marie* The effect of dark brooders on feather pecking on commercial farms
12:15	*Rochais, Céline* Using an appropriate reinforcement for triggering attention and hence performance learning: The example of horse training	*Lambton, Sarah L.* A bespoke management package for loose-housed laying hens: can we mitigate against injurious pecking?
12:30	*Nawroth, Christian* Pigs´ use of human social cues and attentive states in an object choice task	*Nasr, Mohammed Abdel Fattah* The effect of two classes of NSAIDs on the landing ability of laying hens with and without keel fractures
12:45	*Glenk, Lisa Maria* Can dogs relax? Work-related cortisol levels vary in dogs during animal-assisted interventions	*De Jong, Ingrid* Reducing stocking density improves mating behaviour in broiler breeders
13:00	*Haynes, Sally J.* The relationship between human and dog behaviour in animal shelters	*Gebhardt-Henrich, Sabine G.* Individuality of ranging behavior in large flocks of laying hens
13:15–19:00	Excursions	

August 3, 2012 – Friday

08:30–09:10	**Plenary 04:** **Positive emotions** (*Lecture Hall A*)	
	Held, Suzanne & Špinka, Marek Play, positive emotions and animal welfare	
09:15–10:30	**Parallel Sessions** *Lecture Hall A*	*Lecture Hall B*
	Session 06: Positive emotions	***Session 03 continued:*** Intra- and interspecies relationships
09:15	*Seehuus, Birgitte* Behavioural organisation and the reward cycle	*Fels, Michaela* Does origin litter affect social rank of piglets after weaning?
09:30	*Lidfors, Lena* Investigating the reward cycle for play in lamb	*Camerlink, Irene* Behaviour of finishing pigs divergently selected for social genetic effects in barren and straw-enriched pens
09:45	*Reimert, Inonge* Investigating behavioural indicators of positive and negative emotions and emotional contagion in pigs	*Nogueira-Filho, Sérgio L.* The social structure of farmed female capybaras
10:00	*Imfeld-Mueller, Sabrina* Does training for anticipation of a positive reinforcement lead to a long term positive emotional state in growing pigs?	*Murray, Leigh M. A.* Pair-bonding and companion recognition in domestic donkeys (*Equus asinus*)
10:15	***Poster Presentation 07:*** (Positive) emotions and cognition / Physiological parameters Nos. 93-105	***Poster Presentation 08:*** Intra- and interspecies relationships Nos. 81-92
10:30–11:15	**Coffee & posters**	
11:15–12:45	**Parallel Sessions**	
	Session 06 continued: Positive emotions	***Session 03 continued:*** Intra- and interspecies relationships
11:15	*Broom, Donald* What are positive feelings? Examples from achievement excitement, positive cognitive bias and guilt	*Valuska, Annie* Do you smell what I smell? Using olfactory cues to modulate the social behavior of laboratory rabbits
11:30	*Briefer, Elodie F.* Goat kid calls differ according to emotional intensity	*Roth, Beatrice A.* Effects of group stability on aggression, stress and injuries in breeding rabbits
11:45	*Špinka, Marek* Acoustic properties of piglet voices determine what emotional intensity and valence people attribute to them	*Gutmann, Anke K.* Characteristics of social relationships in a dynamic dairy cow herd

12:00	*Verbeek, Else* Morphine induces an optimistic judgement bias after consumption of a palatable food reward in sheep	*Pinheiro Machado Filho, L. Carlos* Influence of social hierarchy on the use of shade by dairy cows
12:15	*Richter, Sophie Helene* A glass full of optimism: enrichment effects on cognitive bias in a rat model of depression	*Tresoldi, Grazyne* Housing effects on social licking in dairy heifers
12:30	*Murphy, Eimear* An active-choice judgement bias task for pigs: a comparison between Göttingen minipigs and conventional farm pigs	*Patt, Antonia* Temporal separation and subsequent reintegration of individual goats: assessment of welfare effects
12:45–14:15	**Lunch**	
14:15–14:55	**Plenary 05:** **Cognition & emotion** (*Lecture Hall A*)	
	Huber, Ludwig The impact of cognitive biology on the question of animal welfare	
15:00–16:00	**Parallel Sessions**	
	Session 07: Cognition and emotion	***Session 08:*** Mutilations
15:00	*Düpjan, Sandra* Design of cognitive bias studies - Review of current research and a pilot study in domestic pigs	*Waiblinger, Susanne* Factors influencing injuries and social stress in horned and dehorned goat herds
15:15	*Franks, Becca* Validating a new behavioural measure of welfare and its relationship to theoretical models of welfare	*Hokkanen, Ann-Helena* Disbudding of calves: the effects of an oral sedative agent on behavioural sedation scores
15:30	*Košťál, Ľubor* Housing conditions and cognitive bias in Japanese quail	*Petherick, Carol* The impact of spaying on the welfare of beef cattle
15:45	*Doyle, Rebecca* Negative stimuli affect the performance of sheep in a spatial cognitive task	*Lomax, Sabrina* Infra-red thermography as a tool for quantitative pain assessment
16:15–17:45	AGM – Annual General Meeting of ISAE	
19:00–23:00	**Conference Dinner at 'Heuriger Wolff'**	

August 4, 2012 – Saturday

09:00–09:40	**Plenary 06:** **Behavioural indicators of welfare** (*Lecture Hall A*)	
	Martins, Catarina Fish welfare: current understanding and future directions	
09:45–11:00	**Parallel Sessions** *Lecture Hall A*	*Lecture Hall B*
	Session 09: Behavioural indicators of welfare	***Session 10:*** Environment, behaviour and welfare II
09:45	*Buckley, L. A.* State-dependent learning: a welfare assessment tool to identify the effects of different diet regimes on hunger state in broiler breeders?	*Stella, Judi* Environmental factors that affect the behavior and welfare of domestic cats (*Felis sylvestris catus*) housed in cages
10:00	*Fureix, Carole* Towards a "depressive-like" state in horses?	*Ellis, Jacklyn J.* Evaluation of environmental enrichment preferences of domestic cats using a choice test
10:15	*Gaskill, Brianna N.* Snug as a bug in a rug: using behavioral thermoregulation to measure thermal stress and improve laboratory mouse welfare	*Cooper, Jonathan* Spatial allowances and locomotion in companion rabbits
10:30	*Berger, Anne* Detection of stress and evaluation of living conditions in any holding system by chronobiological analysis of continuously recorded behavioural data	*Webb, Catherine L.* Effects of citronella and odourless cold air sprays on aversion in dogs
10:45	***Poster Presentation 09:*** Mutilations, pain & welfare / Free Nos. 121-130, 106-107	***Poster Presentation 10:*** Free Nos. 110-120
11:00–11:30	**Coffee & posters**	
11:30–12:30	**Parallel Sessions**	
	Session 09 continued: Behavioural indicators of welfare	***Session 10 continued:*** Environment, behaviour and welfare II
11:30	*Lesimple, Clémence* When reality and subjective perception highly differ: Questionnaire vs objective observations discrepancies in horses' welfare assessment	*Steinmetz, Henriette* The effect of straw length on penmate directed behaviour in pens with growing pigs
11:45	*Kirchner, Marlene K.* Qualitative Behaviour Assessment is independent from other parameters used in the Welfare Quality® assessment system for beef cattle	*Meyer, Susann* On the impact of structural and cognitive enrichment on learning performance, behaviour and physiology of dwarf goats (*Capra hircus*)
12:00	*Dixon, Laura M.* Assessing motivation for appetitive behaviour: food restricted broiler breeders cross a water barrier to access a foraging area without food	*Hickman, Debra* Multilevel caging enhances the welfare of rats as assessed by a spatial cognitive bias assay

12:15	*Ursinus, Winanda W.* Relationship between blood platelet serotonin parameters, brain serotonin turnover and behavioural responses of pigs	*Kistler, Claudia* Preference for structured environment in zebrafish (*Danio rerio*) and checker barbs (*Puntius oligolepis*)
12:30–13:30	**Lunch**	
13:30–14:10	**Plenary 07:** **Environment, behaviour and welfare** (*Lecture Hall A*)	
	Oostindjer, Marije Improving quality of life of newly weaned piglets	
14:15–15:00	**Parallel Sessions**	
	Session 09 continued: Behavioural indicators of welfare	***Session 10 continued:*** Environment, behaviour and welfare II
14:15	*Watters, Jason* The importance of correctly classifying behavior: the case of stereotypic pacing in zoo animals	*Andersen, Inger L.* The "UMB farrowing pen" – a promising system for future loose housing of lactating sows
14:30	*Bravo, Camila* Animal welfare assessment of captive big felids in Chilean zoos using an audit system	*Marchant-Forde, Jeremy N.* Effects of social isolation and environmental enrichment on laboratory housed pigs
14:45	*Koene, Paul* Framework to determine a positive list for mammals suitable as companion animals	*Edwards, Sandra* Optimising nest design for the PigSAFE free farrowing pen
15:00–15:30	**Coffee**	
15:30–16:45	**Parallel Sessions**	
	Session 11: Free papers	***Session 10 continued:*** Environment, behaviour and welfare II
15:30	*Bøe, Knut E.* Thermoregulatory behaviour and use of outdoor yards in dairy goats	*Hötzel, Maria J.* Behaviour response to two-step weaning is diminished in beef calves previously submitted to temporary weaning with nose flaps
15:45	*Hothersall, Becky* Food colour preferences in immature red junglefowl and broiler chickens	*Irrgang, Nora* Can pasture access contribute to reduced agonistic interactions and relaxation in the loose housing barn in horned dairy cows?
16:00	*De Passillé, Anne M.* Weaning dairy calves off milk according to their ability to eat solid feed reduces the negative effects of weaning on energy intake and weight gain.	*Dippel, Sabine* Lying behaviour of primi- and multiparous cows in early lactation
16:15	Vaughan, Alison Cross sucking in milk fed calves may be motivated by a need for oral stimulation and develop into a habit over time	Rajapaksha, Eranda Effect of uncomfortable standing surfaces on restless behavior and muscle activity of dairy cows
16:30	*Webb, Laura* What do calves chose to eat and how do preferences affect calf behaviour and welfare?	*Zerbe, Frank* Cortisol in faeces of fattening bulls: effect of housing conditions, temperature, and behavior
16:45–17:15	Closing	
18:30	Farewell Party at Vetmeduni	

Workshops

Workshop 1: Evaluation of student learning (Lecture Hall D)

Organiser: Cassandra Tucker[1] and Jenny Yngvesson[2]
[1]Department of Animal Science, University of California, Davis, USA
[2]Swedish University of Agricultural Sciences, Skara, Sweden

In a 2-hour workshop, we will provide an overview about pedagogical issues associated with evaluation of undergraduate student coursework, engage in small-group discussion about different methods of evaluation that we currently use in our courses and finish with a presentation about saving time when evaluating students. Our goal with this workshop is to discuss concrete examples and ideas that you can implement in your courses and do this in the light of our experiences and current pedagogical research. Participants should come prepared to discuss specifics of how they currently evaluate undergraduate student coursework, including the methods of evaluation (exams, written assignments, oral exams etc.), topics covered, number of students, their level of skill, time required, and any advantages/disadvantages of your examples. Once registered, participants will be sent a worksheet or URL link to complete before the workshop to facilitate a summary of evaluation methods used that will be made available to all.

This workshop will be complemented by a short oral presentation by Anna Olsson on the involvement of students in animal welfare blogging (page 106).

Workshop 2: Farm animal welfare research and education in an enlarged Europe and beyond (Lecture Hall M)

Organiser: Marek Špinka[1], Ľubor Košťál[2], Gudrun Illmann[1], Alain Boissy[3], Fritha Langford[4], Christoph Winckler[5], Boris Bilčík[2], Marlene Kirchner[5], Michala Melišová[1], Linda Keeling[6]
[1]Research Institute for Animal Production, Prague, Czech Republic
[2]Institute of Animal Biochemistry and Genetics SASci, Ivanka pri Dunaji, Slovakia
[3]INRA UMR 1213 Herbivores, F-63122 Saint-Genès-Champanelle, France
[4] SAC, Easter Bush, Midlothian, *UK*
[5]University of Natural Resources and Life Sciences, Vienna, Austria
[6]University of Agricultural Sciences, Uppsala, Sweden

The aim of the workshop is to present the preliminary results and forthcoming activities of the EU funded project entitled Animal Welfare Research in an Enlarged Europe (acronym: AWARE, www.aware-welfare.eu). Overall goal of AWARE is to promote integration and increase the impact of European research on farm animal welfare through the development of Europe-wide networks of scientists, lecturers and students, and by establishing a network of stakeholders active in farm animal welfare (FAW) knowledge transfer and implementation.

AWARE is organized in four mutually supportive Work Packages. WP1 enhances the integration of farm animal welfare research by fostering collaboration based on mutual recognition and by enhancing networking and proposal writing skills in motivated researchers throughout the enlarged Europe. WP2 promotes networking in farm animal welfare university education, thus enhancing opportunities for young scientists in new and candidate countries to start research in field of farm animal welfare. WP3 focuses on enhancing public awareness, promoting implementation of EU policies, and facilitating uptake of and facilitating uptake of farm animal welfare research. WP4 facilitates mobility of researchers and students.

The workshop will use the advantage of the ample attendance of the ISAE congresses and the suitable geographic location of Vienna (four countries neighbouring Austria are the new member countries of the EU that together with candidate countries are in the centre of interest of AWARE) for the presentation of the results obtained during the first year of the project on mapping of farm animal welfare research and education in an enlarged Europe and for the presentation of the future plans. Two other EU funded FAW projects ANIHWA (www.anihwa.eu) and AWIN (www.animal-welfare-indicators.net) will be presented and the workshop will be used for an exchange of information between these projects on the topics of mutual interest (mainly mapping farm animal welfare research with ANIHWA and mapping global animal welfare education with AWIN).

Within the general discussion participants of the workshop will address the questions: i) how to improve the involvement of the new member and candidate countries of the EU in European FAW research, and ii) how to increase the European research capacity in FAW through integrating the underutilized human and knowledge potential in the new and candidate countries.

Table of contents

David Wood-Gush Memorial Lecture

Plenary presentations

Session 01. Behavioural development: genes, epigenetics and ontogeny

Session 02. Temperament/Individual differences

Theatre **Session 02 no. Page**

Session 03. Intra- and interspecies relationships

Theatre **Session 03 no. Page**

Session 04. Behavioural methods

Theatre **Session 04 no.** **Page**

Session 05. Environment, behaviour and welfare I

Theatre **Session 05 no.** **Page**

Session 06. Positive emotions

Session 07. Cognition and emotion

Session 08. Mutilations

Session 09. Behavioural indicators of welfare

Session 10. Environment, behaviour and welfare II

Session 11. Free papers

Theatre **Session 11 no.** **Page**

Workshop 1

Poster session

The social modulation of behaviour during development

Sachser, Norbert, University of Muenster, Department of Behavioural Biology, Badestrasse 13, 48149 Münster, Germany; sachser@uni-muenster.de

Individual differences in cognition, emotion, and behaviour develop during ontogeny. Using a comparative approach, this contribution focuses on the role of the social environment during early phases of life, during adolescence, and in adulthood for the modulation of behavioural profile. For gregarious species, the stability of the social environment in which the pregnant and lactating female lives is of major importance for foetal brain development and the behavioural profile of the offspring in later life. Social instability during these critical periods of development generally brings about a behavioural and neuroendocrine masculinisation in daughters and a less pronounced expression of male-typical traits in sons. Moreover, when mothers live in a socially challenging world during pregnancy and lactation, emotional behaviour of their offspring can be intensified in adulthood. These effects of the social environment are likely to be mediated by maternal hormones and/or maternal behaviour. From an evolutionary perspective, the behavioural effects of social experiences during these early phases of life are not necessarily 'pathological' (nonadaptive) consequences or constraints of adverse social conditions. Rather, mothers could be adjusting the offspring to their environment in an adaptive way. It is of note, however, that the susceptibility to the maternal effects is significantly influenced by the offspring's genotype. Adolescence is another period in which behavioural development is particularly susceptible to social influences. There is some evidence that social experiences at this time alter and canalize behaviour and endocrine stress responses in an adaptive fashion, so that earlier influences on behavioural profile development can be complemented and readjusted, if necessary, to meet current environmental conditions. In terms of underlying neuroendocrine mechanisms, a central role for the interaction of testosterone and stress hormones is suggested. Interestingly, throughout the whole life-time, contact with social companions can buffer the behavioural and hormonal stress response in challenging situations. During the early postnatal phase usually the mother functions as a secure base for the infant. In adulthood the occurrence of social buffering critically depends on the nature of the relationship between the individuals involved. In summary, the social modulation of behavioural profiles and stress responses, which can occur in the time span from the prenatal phase to adulthood, seems to be an effective mechanism for repeated and rapid adaptation.

Behaviour epigenetics – the connection between environment, stress and welfare

Jensen, Per, Nätt, Daniel and Goerlich, Vivian, IFM Biology, Linköping University, Avian Behavioural Genomics and Physiology group, Linköping University, 58183 Linköping, Sweden; perje@ifm.liu.se

It is generally believed that evolution and selection works on random mutations only, whereas the environment is not thought to directly influence the genome. However, the field of epigenetics suggests mechanisms, whereby, for example, stress can alter the expression of certain genes by means of DNA-modifications (for example, methylation of cytosine), which do not affect DNA sequences. These effects could in principle be inherited, opening a novel way for populations to adapt. It is not clear if behaviour in turn can be affected by epigenetic variation. We studied this in a series of experiments, aiming at understanding possible mechanisms, and their relevance from a welfare perspective. In three separate studies, we found that chronic stressful conditions, as well as brief stress during early life, caused heritable changes in brain gene expression in chickens. This was associated with several behaviour changes in both affected animals and their offspring, and alterations of corticosterone responses to acute stress (stressed vs. controls: $P<0.05$; repeated measures ANOVA). Furthermore, domesticated White Leghorn chickens showed a significantly different methylation pattern than Red Junglefowl in 821 out of 3623 arbitrarily chosen genes. The differential methylation was heritable, and stable in up to eight generations. White Leghorns were hypermethylated in 79% of them ($\chi 2=49.8$, $P<0.0001$), indicating that they have acquired many novel methylations during domestication. We suggest that epigenetic mechanisms mediate stress effects on behaviour and welfare in a heritable and adaptive manner. Hence, environmental effects can cause long-lasting modifications of the orchestration of the genome.

The role of genes, epigenetics and ontogeny in behavioural development

Rodenburg, Bas, Wageningen University, Animal Breeding & Genomics Centre, P.O. Box 338, 6700 AH Wageningen, the Netherlands; bas.rodenburg@wur.nl

The behavioural characteristics of an individual are determined by its genes and by its physical and social environment. Not only the individual's early-life and current environment is of importance, but also the environment of previous generations. Through epigenetic processes, stress in parents and even grandparents can translate in changes in behavioural and physical characteristics of the offspring. Epigenetic processes cause changes in gene function without a change in the gene sequence, by changing the folding of the chromosomes and affecting gene expression. Apart from epigenetics, also maternal hormones excreted prior to egg-laying or during pregnancy have effects on behavioural development of the offspring. The environment during ontogeny has considerable impact on behavioural development. Absence or presence of maternal care has been shown to have strong effects: in laying hens, we found that maternal care resulted in birds that were less fearful and developed less damaging behaviour when they were adult. Further, changes in the serotonergic system were detected, which plays a role in coping with fear and stress. This relationship between fearfulness, feather pecking and the serotonergic system could be confirmed in a recent genetic comparison of six brown and six white pure-bred genetic lines of laying hens. When studying the genetic profile of these lines, a mutation in the HTR2C gene, a receptor gene of the serotonergic system, was detected that was associated with feather damage. This mutation was present in 84% of the white birds and in only 35% of the brown birds, which corresponds with the differences in feather pecking and damage between the lines. Genomic information as well as gene expression studies offer perspective for a better understanding of the role of genes, epigenetics and ontogeny in behavioural development. A favourable environment, both during ontogeny and during later life, plays a central role in behavioural development.

Social attachment in sheep: similarities and differences between intra- and interspecies relationships

Nowak, Raymond[1] and Boivin, Xavier[2], [1]CNRS-INRA, Animal Physiology, Health and Husbandry, UMR Physiologie de la Reproduction et des Comportements, 37380 Nouzilly, France, [2]INRA, Animal Physiology, Health and Husbandry, UMR1213 Herbivores, 63122 Saint-Genès Champanelle, France; raymond.nowak@tours.inra.fr

Farm animals develop an important social network with intra- and interspecific partners, including humans. In some cases this can lead to strong emotional bonds indicating the existence of attachment. There is no doubt that a better understanding of the mechanisms of attachment will benefit animal welfare. As a model, the sheep is well known to develop various forms of social attachment (mothers towards young, lambs towards siblings). The relationship they can develop with humans is much less understood. In this review we outline features and mechanisms that participate in the development and the expression of intra- and interspecific social attachment in lambs. Behavioural tests comparing responses towards a presumed attachment figure with those directed towards unfamiliar or familiar conspecifics demonstrate that lambs do search specifically the proximity of their mother, sibling or human caretaker. Differential emotional responses in the presence (calmness) or the absence of the partner (agitation) are also expressed. However, a relationship with a human takes place more easily when lambs are reared without their primary attachment figure, the mother. Human-lamb attachment is then facilitated by positive social contacts (stroking) provided by a specific caretaker. In the case of attachment with the mother, suckling is the main reward. Positive social contacts including stroking or suckling activate similar brain regions and trigger a release of oxytocin, a neuropeptide known to play a key role in prosocial behaviours. Recent evidence even supports a role for oxytocin in the development of filial attachment. In conclusion, lambs develop intra- and interspecific attachment but not in a concomitant manner as the presence of the mother strongly reduces the lamb's motivation to interact with a human. However, under artificial rearing conditions the human becomes a salient attachment figure which may facilitate the management of lambs on farm.

Play, positive emotions and animal welfare

Held, Suzanne[1] and Špinka, Marek[2], [1]Animal Welfare and Behaviour Group, School of Veterinary Science, University of Bristol, Langford BS40 5DU, United Kingdom, [2]Institute of Animal Science, Department of Ethology, Přátelství 815, 104 00 Prague – Uhříněves, Czech Republic; suzanne.held@bris.ac.uk

Play has long been identified as a potential welfare indicator: it disappears when animals are under fitness challenge and is thought to be accompanied by a pleasurable emotional experience. But animal play is a tricky behavioural phenomenon, characteristically flexible and variable within and between species, with its proximate mechanisms and ultimate functions still not fully understood. Its relationship to animal welfare is therefore complex and merits a focused theoretical investigation. The talk will review evidence on three aspects of the play-welfare relationship: its utility as a welfare indicator, its biological function (or functions?) and its proposed effects on an animal's welfare. Within this framework it considers whether play always indicates the absence of fitness threats, whether and how it indicates the presence of pleasurable emotional experiences; whether it brings immediate psychological and long-term fitness and health benefits, and thus improves current and future welfare; and whether it is socially contagious and therefore can spread good welfare in groups. The evidence suggests that play does indeed hold promise as a welfare indicator and also as a tool to improve it. However, it also points to difficulties in its study and interpretation, and raises some unresolved questions. A better fundamental understanding is needed of the varied ultimate functions and proximate mechanisms of play, and the species-specific play patterns of captive animals, so that we may explain exactly what an animal's play behaviour tells us about its welfare state, and whether and how play might be applied as a tool to improve welfare.

The impact of cognitive biology on the question of animal welfare

Huber, Ludwig, Comparative Cognition, Messerli Research Institute, University of Veterinary Medicine Vienna, Veterinaerplatz 1, 1210 Vienna, Austria; Ludwig.Huber@vetmeduni.ac.at

The aim of this talk is to set a common stage for our knowledge of socio-cognitive abilities of non-human animals for the following ethical considerations by philosophers and people concerned with animal welfare. Although there is an increasing understanding for the need to look more closely into the growing field of empirical research of cognition and emotion in non-human animals, not only of our closest relatives, primates, but also into more distant branches of the 'tree of life', the philosophical and ethical discussion is hampered by the difficulty to distinguish between solid and suggestive or even anecdotal evidence of how non-human animals view and understand their physical and social environment. Of course, reports of surprisingly 'clever' performances of non-human animals in the mass media have significantly contributed to the people's appreciation and acceptance of these creatures as intentional and sentient beings. But still there is the risk of over-interpretation and premature conclusions, which causes unwarranted resistance by philosophers and lawyers. In this brief paper I shall therefore focus on those examples from the current research on cognitive abilities in the physical and social domain, where empirical evidence is solid and conclusive: tool innovation and perspective taking.

Fish welfare: current understanding and future directions

Martins, Catarina, Centro de Ciências do Mar (CCMAR), Aquaculture, Campus de Gambelas, 8005-139 Faro, Portugal; cimartins@ualg.pt

Fish farming is the fastest growing sector in world production of animal-derived food. The average annual growth rate of aquaculture is 8.3% as compared to only 2.7% in poultry and livestock. Already this year (2012), 50% of the fish consumed by humans will come from aquaculture. Not only the number of farmed fish is increasing but also farming systems are becoming more intensive. At the same time, there is evidence that fish are sentient creatures and therefore capable of experiencing good and poor welfare. In this presentation, the main cognitive, neuroanatomical and emotional evidences supporting the importance of addressing welfare in farmed fish will be reviewed. The most relevant welfare issues within the aquaculture industry will be discussed including poor water quality, impaired health, high stocking densities, handling, grading and slaughter methods. In addition, the main welfare indicators will be reviewed with especial emphasis on operational behavioural indicators. Changes in foraging behaviour, ventilatory activity, aggression, individual and group swimming behaviour, stereotypic and abnormal behaviour have been linked with acute and chronic stressors in aquaculture and can therefore be regarded as likely indicators of poor welfare. On the contrary, measurements of exploratory behaviour, feed-anticipatory activity and reward-related operant behaviour are beginning to be considered as indicators of positive emotions and welfare in fish. Future research directions in the field of fish welfare will be discussed. These will include the incorporation of concepts such as allostasis, coping strategies and cognitive appraisal in fish farming.

Improving quality of life of newly weaned piglets

Oostindjer, Marije[1], Bolhuis, J. Elizabeth[2], Van Den Brand, Henry[2] and Kemp, Bas[2], [1]Norwegian University of Life Sciences, Chemistry, Biotechnology and Food Science, P.O. Box 5003, 1432 Ås, Norway, [2]Wageningen University, Animal Science, Adaptation Physiology group, P.O. Box 338, 6700 AH Wageningen, the Netherlands; marije.oostindjer@umb.no

The growing concern of the general public regarding food safety and animal welfare results in a desire for a reduction in antibiotic use and a more sustainable animal production. In pigs, the period around weaning tops the other periods of a pig's life when it comes to antibiotic use in the Netherlands. Difficulties to adapt to the postweaning situation and its many stressors are one of the causes of a low feed intake after weaning and associated intestinal problems. This presentation provides a synthesis of our work from the past years on improving early solid feed intake and consequently the ability of piglets to adapt to the postweaning situation, by allowing piglets to learn more from their mother in the preweaning period. All statistics were done with (repeated) mixed models including treatment and batch, on pen level. We show that providing piglets with the possibility to eat together with their mother and to interact more with their mother in general not only increases early solid feed intake but also increases behavioural indicators of welfare postweaning such as play, and may act as a buffer to a loss of reward after weaning. Enrichment before weaning also increases solid feed intake around weaning, while postweaning enrichment results in a better growth, less intestinal problems, more play and less damaging behaviour. Finally, we show that learning from flavour cues in the maternal diet before birth and during lactation reduces weaning stress when this familiar flavour is present in the postweaning environment, and this reduction results in increased feed intake, growth and play behavior, and reduced damaging behaviour. We combine results from these three pathways of learning to provide a synthesis on how to improve the quality of life of the newly weaned pig, which is an important step towards sustainable, antibiotic-low pig production.

Prenatal undernutrition affects spatial memory in male lambs in a genotype-dependent manner

Dwyer, Cathy M.[1], Mcilvaney, Kirstin A.[1], Erhard, Hans W.[2], Rooke, John A.[1] and Ashworth, Cheryl J.[3], [1]SAC, Animal Behaviour and Welfare, Roslin Institute Building, EH25 9RG, Edinburgh, United Kingdom, [2]AgroParis Tech, 16 rue Claude Bernard, 75231 Paris Cedex 05, France, [3]The Roslin Institute, Developmental biology, The Roslin Institute Building, Easter Bush, EH25 9RG, Midlothian, United Kingdom; cathy.dwyer@sac.ac.uk

We have shown that moderate prenatal undernutrition in the first two thirds of gestation affects lamb reactivity and birth weight in a breed-dependent manner. This study investigated whether this treatment affected spatial memory, side preference and exploratory behaviour of juvenile male lambs. Forty 7-month old entire male lambs of two breeds adapted to a harsh environment (Scottish Blackface, BF) or to more benign conditions (Suffolk, S) were used. The dams of these lambs had either been undernourished at 75% requirements from d1-90 gestation (R: BFR=8, SR=13) or fed 100% throughout (C: BRC=8, SC=11). Lambs were tested in mazes for spatial memory ability, side preference, and willingness to leave their group to explore a novel arena/objects. The effects of litter size at birth, genotype and prenatal nutritional treatment were tested with multivariate statistics or Kruskal-Wallis tests. 87.5% of lambs showed a right-side bias, there were no effects of breed or treatment. Lambs took a similar time to navigate the maze on the memory acquisition trial, but the SR lambs were slower than SC, and BFR quicker than BFC, to exit the maze in the memory retention trial (median time (s): BFC=44.0, BFR=27.0, SC=42.0, SR=68.0; breed × nutrition: Wald=4.15, $P<0.05$), and this tended to influence the number of errors made ($P=0.10$). BF lambs spent longer walking in the novel arena ($P<0.05$) and less time in the area next to other lambs than S lambs (mean time (s): BF=168.8, S=231.8 s.e.d.=26.8, $P<0.05$), but there were no effects of treatment. There was also no effect of breed or treatment on the number of novel objects lambs approached, or the time spent interacting with objects. The data suggest that moderate prenatal undernutrition in the first 90 days of gestation had relatively few effects on juvenile lamb behaviour, but seemed to affect spatial memory in a breed-dependent manner.

Association between ewe undernutrition around conception and laterality and emotional responsiveness in the offspring

Matthews, Lindsay R.[1], Hernandez, Carlos E. [2], Harding, Jane E.[3], Oliver, Mark H.[3] and Bloomfield, Frank H.[3], [1]University of Auckland, Psychology, Private Bag 92019, Auckland 1142, New Zealand, [2]University of New England, School of Environmental and Rural Sciences, University of New England, Armidale, NSW 2351, Australia, [3]University of Auckland, Liggins Institute, Private Bag 92019, Auckland 1142, New Zealand; lindsay.matthews1@gmail.com

Emotional responses are controlled predominantly by the right hemisphere and left lateralised mammals express heightened emotional responses. A variety of genetic and developmental factors contribute to brain lateralisation during early development. This study examined, for the first time, the influence of modest periconceptual undernutrition on brain asymmetry and emotional responsiveness. Ewes were well-fed (Controls) or mildly underfed (UN)(10-15% weight reduction) for 60 d before mating UN-60), 30 d after mating (UN+30) or both before and after mating (UN90). Lamb behavioural lateralisation (right-left preference in a y-maze; n=87 at 4 mo, n=67 at 18 mo) was assessed at 4 and 18 mo of age, and their behavioural (escape attempts) and plasma cortisol responses to social isolation at these ages (n=130 at 4 mo, n=79 at 18 mo) were also measured. Data were analysed by Fisher's exact test, Restricted Maximum Likelihood analysis or as Poisson variates using generalised linear model. In response to isolation, lambs born to the underfed ewes showed fewer escape responses than controls at 4 mo of age (Control=0.4±0.1, UN-60=0±0, UN+30=0.2±0.2, UN90=0.2±0.1 attempts to escape, P<0.001), and offspring from UN90 had significantly reduced cortisol (area under the curve) at 18 mo (Control=546.5±46.8, UN−60=427.2±67.7, UN+30=448.5±249.4, UN90=438±42.2 ng/ml min, P<0.05). The strong left-side bias of control male lambs was absent in undernourished (UN90) males (100% vs. 44% turning left, P<0.05) and strong right bias of females tended to neutrality. The reduced left-right brain asymmetry and behavioural and physiological responses to stressful situations are indicative of blunted emotional responsiveness in lambs born to mildly underfed ewes. This study provides evidence that the dam's nutritional status around conception is associated with brain lateralisation and coping style.

Impacts of prenatal stress and environmental enrichment during lactation on the behaviour and performances of piglets after weaning

Ringgenberg, Nadine[1,2], Bergeron, Renée[3], Torrey, Stephanie[1,2] and Devillers, Nicolas[2], [1]University of Guelph, Animal and Poultry Science, Guelph, ON, N1G 2W1, Canada, [2]Agriculture and Agri-Food Canada, Dairy and Swine R & D Centre, 2000 College St, Sherbrooke, QC, J1M 0C8, Canada, [3]University of Guelph, Alfred Campus, 31, rue St-Paul – C.P. 580, Alfred, ON, K0B 1A0, Canada; nringgen@uoguelph.ca

The impact of prenatal stress and enrichment in lactation on the behaviour of 246 weaned piglets from 41 litters was studied in a 2×2 factorial experiment. Sows were assigned to either a social stress treatment (T) during mid-gestation or a control group (C). In lactation, sows were either housed in straw enriched pens (E) or in standard farrowing crates (S). Six piglets per litter were weaned without mixing at 21 days of age (D1) in standard tenderfoot pens. Their behaviour was video recorded for 6 h per day on D1, D2 and D7. Resting and feeding were measured using 5-min scan sampling and social behaviours were observed using 1-min one-zero sampling. Growth and feed intake were also recorded. Data were analysed using the GLIMMIX and MIXED procedures of SAS with the group as experimental unit. Compared to C piglets, T piglets showed less aggression (17±2 vs. 24±2%; $P<0.05$), nosing-chewing on littermates (43±3 vs. 52±3%; $P<0.1$), playing (7±2 vs. 12±2%; $P<0.05$), mounting (5±1 vs. 11±1%; $P<0.05$) and more resting (63±3 vs. 54±3%; $P<0.05$) and belly nosing (13±2 vs. 9±2%; $P<0.1$) on D7. Compared to S piglets, E piglets showed less overall aggression (10±2 vs. 20±2%; $P<0.05$) and playing (3±1 vs. 12±2%; $P<0.05$), less nosing-chewing on D1 (26±3 vs. 36±3%; $P<0.05$) and more resting (62±3 vs. 55±3%; $P<0.1$) and belly nosing (14±2 vs. 8±2%; $P<0.05$) on D7. E piglets lost less weight (-77±47 vs. -228±48 g/piglet; $P<0.05$) and consumed more feed (650±62 vs. 484±64 g/group; $P<0.1$) between D1 and D3 than S piglets. Prenatal stress and enrichment in lactation had negative effects on social behaviour with less natural (play, nosing-chewing, mounting) and more undesirable (belly-nosing) behaviours. Although the transition from an enriched to a barren environment appears to have been stressful, it also led to a better feeding transition after weaning.

Domestication effects on stress recovery in chickens (*Gallus gallus*)

Ericsson, Maria and Jensen, Per, Linköping University, Biology, IFM Biology, Linköping University, 581 83 Linköping, Sweden; miaer@ifm.liu.se

Domestication has seemingly increased the stress tolerance in animals, probably as a result of adaptation to the stressors the domestic animals face. It is possible that this is also true for the ability to recover from an acute stress experience. The aim of the study was two-fold: (1) to develop a method for measuring behavioural stress recovery; and (2) to compare recovery in a domesticated chicken and the wild ancestor. Males and females from two chicken breeds were studied, the domestic layer White leghorn (WL) and the wild progenitor to domestic chickens, the Red Junglefowl (RJF). Following 24 h of habituation to an enriched test arena, in which birds were kept on their own with a mirror as social stimulation, baseline behaviour was measured for one hour before (baseline) and after (recovery phase) an acute stress exposure (restraint) for 3 minutes. A GLM analysis showed significant breed effect on foraging behaviour. The WL birds showed a larger decrease of foraging time ($P=0.04$) and performed fewer ground pecks ($P=0.009$) after stress, than the RJF. WL males also vocalized more than RJF males post stress ($P=0.04$). The behaviour differences indicates that the WL is more affected by stress evoked by restraint than RJF. From a welfare perspective, the results indicate that stress coping should be accounted for when breeding animals for production purposes.

Long term effects of mother rearing: challenge responses of primiparous dairy cows

Wagner, Kathrin[1], Seitner, Daniel[1], Barth, Kerstin[2], Palme, Rupert[3] and Waiblinger, Susanne[1], [1]Institute of Animal Husbandry and Welfare, University of Veterinary Medicine, Vienna, Veterinärplatz 1, 1210 Vienna, Austria, [2]Institute of Organic Farming, Johann Heinrich von Thünen-Institute, Federal Research Institute for Rural Areas, Forestry and Fisheries, Trenthorst 32, 23847 Westerau, Germany, [3]Institute of Biochemistry, University of Veterinary Medicine, Veterinärplatz 1, 1210 Vienna, Austria; Kathrin.Wagner@vetmeduni.ac.at

We investigated the effects of mother vs. artificial rearing on response to social and non-social challenges in adulthood. Rearing of treatment groups only differed the first 12 weeks of life. Control animals were separated from their mother within 24 h after birth and fed via an automatic milk feeder (n=11; two or six times a day). They were housed in one group together with animals suckled by their mother (Mother, n=15; twice a day 15 min in the calf area or calves had permanent access to mother and cow herd via selection gates). Animals were integrated into the loose housed cow herd two months before calving. About 4.5 months after their first calving (age 31±1.4 months) animals were confronted with a novel object two times (first: pylon, second: ball) in the alley where animals walked back after milking, and one isolation test. Data were analysed using Mann-Whitney-U-test. With the pylon, Mother-animals less often walked in the middle of the alley (median, min-max: number of middle squares entered; Mother: 6.5, 3.0-8.0; Control: 8.0, 5.0-8.0; $P<0.05$) and tended to walk more at the side of the alley further away from the pylon (Mother: 3.0, 1.0-6.0; Control: 2.0, 0.0-4.0; $P<0.1$). With the ball, Mother stood longer (8.6 s, 1.0-92.3; Control: 3.7 s, 0.0-10.0; $P<0.05$), tended to have shorter latency to stop walking and higher frequency of standing/walking. During isolation, duration of head position 'normal' was shorter in Mother (145.3 s, 91.3-460.2; Control: 272.7 s, 121.3-515.6; $P<0.05$); they tended to walk longer. Difference to baseline of saliva cortisol after 15 min isolation tended to be lower in Mother (0.2 ng/g, -0.7-0.9) than in Control (0.3 ng/g, 0.1-0.8; $P<0.1$). Mother reared animals were more active and attentive to the environment and physiological stress responses during isolation were lower. Mother rearing had long-term effects on challenge response in 2.5 year old cows.

The effect of post-natal environmental enrichment on tail biting in growing pigs

Telkänranta, Helena[1,2], Swan, Kirsi[2], Hirvonen, Heikki[1] and Valros, Anna[2], [1]University of Helsinki, Department of Biosciences, P.O. Box 56, 00014 University of Helsinki, Finland, [2]University of Helsinki, Department of Production Animal Medicine, P.O. Box 57, 00014 University of Helsinki, Finland; helena.telkanranta@helsinki.fi

Redirected explorative behaviour has been proposed as a contributing factor to tail biting in pigs. We hypothesized that providing material for oral manipulation from birth to weaning reduces oral manipulation of other piglets and later tail-biting. We allocated 59 sows and their litters to two treatments in a matched-pairs design. Farrowing pens in the control treatment (n=29) had enrichment representing the minimum required by Finnish legislation: wood shavings twice a day and one suspended plastic ball. Farrowing pens in the enriched treatment (n=30) had the same plus newspaper (non-glossy, partially made of recycled fibre) twice daily and 10 suspended sisal ropes. All piglets were undocked. The piglets were weaned during week 4, and two litters were combined in each growing pen, all of which had the same level of enrichment. Targets and frequencies of oral manipulation were recorded on video during weeks 2, 3, and 9. Tail damage was scored during week 9. Weaning weight and age did not differ between the treatment groups. Oral manipulation of pen-mates differed between treatments: in control pens, the mean numbers of manipulations/piglet/minute were 0.83 and 0.82 at weeks 2 and 3 respectively; in enriched pens, the means were 0.60 and 0.56 at weeks 2 and 3 respectively ($P<0.001$, repeated-measures ANOVA, PASW 18). At week 9, no significant difference was found in the observed behaviour parameters. The prevalence of tail damage of the worst category 4 (part of tail missing) was 8.6% in the control treatment and 0.9% in the enriched treatment ($P<0.001$, Fisher's exact, Statistix 9). The prevalence of category 3 (wounds with inflammation) was 23.5% in the control treatment and 8.8% in the enriched treatment ($P<0.001$, G-test, Statistix 9). It is concluded that additional enrichment in early life has promising implications for reducing tail biting.

The influence of neonatal environment on piglet play behaviour and post-weaning social and cognitive development

Hopkins, Jessica E.[1,2], Ison, Sarah H.[2] and Baxter, Emma M.[2], [1]The University of Edinburgh, Royal (Dick) School of Veterinary Studies, Easter Bush Veterinary Centre, EH25 9RG, United Kingdom, [2]Scottish Agricultural College, Animal and Veterinary Sciences Research Group, SAC, West Mains Road, Edinburgh, EH26 0PH, United Kingdom; Jessica.Hopkins@sac.ac.uk

Domestic pigs are highly playful animals throughout development. The function of play is not fully understood, but current theories suggest it is influential in successful socio-cognitive development, particularly during neonatal stages. Therefore commercial neonatal environments (NE) are fundamental to successful stimulation and development of play in neonatal pigs. Substantial indirect and direct socio-cognitive effects post-weaning could influence social interactions necessary in commercial environments that raise welfare concerns (e.g. mixing aggression). This study investigated how a) piglet play developed in two pre-weaning NEs; the farrowing crate (NEC) and an enriched loose-housed environment (PigSAFE pen (NEP)); b) play developed post-weaning in standard commercial weaner pens and c) effects on cognitive abilities in Spontaneous Novel Object Recognition Tests post-weaning. Hourly focal sampling recorded play behaviours pre- and post-weaning in 72 mixed-sex piglets (36 per NE), from 12 litters (6 per NE), between 08:00 and 16:00, three days per week. 24 of the piglets underwent cognitive tests five weeks post-weaning, involving timing interactions with familiar and novel objects after two retention times (RTs) (15 and 60 minutes) and initial exposure to familiar objects. REML analyses showed NE had a significant affect on play pre-weaning; with NEP piglets performing more varied and higher amounts of play behaviour ($F=5.06_{(1,71)}$, $P=0.051$). There was no effect post-weaning. For cognitive tests, NEP piglets spent more time interacting with the novel object during RT-15 compared to NEC piglets ($F=5.39_{(1,23)}$, $P=0.045$). There was no NE effect for RT-60. It was concluded that NE is fundamental to stimulating play, resulting in successful socio-cognitive development, relating to play function theories of training for the unexpected. However, its effects on play are short-term and highly dependent on the present environmental stimulus, suggesting that any life-long benefits play may have on an animal's welfare can only be achieved by regular environmental stimulation through-out life (e.g. constant enrichment).

Temperament dependent reactivity of calves in challenging test situations: a multivariate and correlational analysis of behavioural and physiological measures

Graunke, Katharina L.[1], Nürnberg, Gerd[2], Schön, Peter-Christian[1] and Langbein, Jan[1], [1]Leibniz Institute for Farm Animal Biology (FBN), Research Unit Behavioural Physiology, Wilhelm-Stahl-Allee 2, 18196 Dummerstorf, Germany, [2]Leibniz Institute for Farm Animal Biology (FBN), Research Unit Genetics and Biometry, Wilhelm-Stahl-Allee 2, 18196 Dummerstorf, Germany; graunke@fbn-dummerstorf.de

Original research has often described temperament in animals one-dimensional (proactive-reactive, aggressive-non-aggressive, etc.), while theoretical papers mostly argue for two or more dimensions. The aim of this study was to develop at least two-dimensional temperament types (TT) and describe them by analysing supposedly consistent behaviour in two standard behaviour tests and simultaneous measurement of heart rate variability (HRV). We tested 361 calves (175 male, 186 female) of Holstein Friesian × Charolais crossbreed at 90 dpp±3 d in an open field (4.0×9.5 m). They were fitted belts for HRV-measurements (Polar Electro, Finland). Test protocol allowed for 10 min acclimatisation to the open field, 10 min novel-object-test (pylon) and 10 min novel-human-test (unknown human). Behaviour was live-recorded using The Observer 5.0 (Noldus, The Netherlands). Behavioural variables were correlated using Spearman rank correlation, standardised, centred and analysed by a factor analysis (SAS 9.2, SAS Institute Inc., USA) separately for each test. Scores for each animal in each factor were calculated from original data and the respective loadings and animals were classified into TT accordingly. The influence of TT, sex and their interaction on the RMSSD/SDNN ratio (HRV) was analysed by a generalized linear mixed model. Variation in the data was best described by two factors (Velicer's Average Square Correlation smallest). Calves scoring within a circle of 1 SD around the mean score were classified as 'inconclusive'; the others according to the quadrant they scored in. Neither sex nor interaction of sex × TT influenced the RMSSD/SDNN ratio, but TT did ($F=7.76$, $P<0.001$). Two TT had a higher ratio in both tests than during baseline measurement and differed significantly ($P=0.026$-$P<0.001$) from the two other TT having a lower ratio. According to behaviours important for the factors and the change of the HRV, TT could be described as 'stressed-fearful', 'excited-interested', 'calm-indifferent', 'relaxed-outgoing'.

Linking cortisol responsiveness and aggressive behaviour in gilthead seabream *Sparus aurata*: evidence of divergent coping styles

Castanheira, Maria F.[1], Herrera, Marcelino[1,2], Costas, Benjamín[3], Conceição, Luís E.C.[1] and Martins, Catarina I.M.[1], [1]Centre of Marine Sciences CCMAR, Aquaculture Research Group, Universidade do Algarve, Campus de Gambelas, 8005-139 Faro, Portugal, [2]IFAPA, Agua del Pino. Cra. Cartaya-P. Umbría, 21450 Cartaya, Spain, [3]Centro Interdisciplinar de Investigação Marinha e Ambiental (CIIMAR), Universidade do Porto, 4050-123 Porto, Portugal; mcastanheira@ualg.pt

Farmed animals often exhibit a pronounced individual variation in stress coping strategies. Farmed fish also have been shown to exhibit divergent coping styles with proactive and reactive individuals differing in a variety of neuroendocrine and behavioural responses. In this study we disclose for the first time that individual differences in cortisol responsiveness after a restraining test are predictive of aggressive behaviour in seabream Sparus aurata, one of the most important European farmed fish. Seabream juveniles (n=24, initial weight: 49.31±7.25 g) were exposed to a restraining test that consisted of keeping each fish in an emerged net for three minutes. Afterwards fish returned to their home tank and 30 min after were rapidly caught and anaesthetised with 2-phenoxyethanol prior to blood sampling. Blood was collected from the caudal vein using heparinised syringes and analysed for cortisol (radioimmunoassay). After 3 months the same individuals were exposed to an aggression test: for 15 min fish were allowed to interact with a naive fish of similar size (max 10% weight difference). Aggressive behaviour was determined (latency to start chasing: time taken until the first chase; number of chases: sudden change in swimming direction and speed as a response to an approach by the opponent) and correlated with cortisol responsiveness. During the aggression test, no fish showed any signs of physical injury. Results show that individuals exhibiting lower cortisol responsiveness after a restraining test are more aggressive (r_s=0.503, P=0.014 cortisol vs. latency to chase; r_s=-0.479, P=0.021 cortisol vs. number of chases). These results are in agreement with previous studies using lines of fish with divergent coping styles. In conclusion, juveniles of seabream exhibit pronounced individual differences in cortisol responsiveness and aggression that are related and likely to be distinctive traits of coping styles.

Perseverative parrots: individual differences and changes in perseveration in orange-winged Amazon parrots (*Amazona amazonica*)

Cussen, Victoria and Mench, Joy, University of California, Davis, Animal Biology Graduate Group and Center for Animal Welfare, One Shields Avenue, Davis, CA 95616, USA; vacussen@ucdavis.edu

An increase in perseveration (generalized, uncontrollable response repetition) is hypothesized to underpin stereotypies, and high rates of perseveration have been found in *Amazona amazonica* with established stereotypies. We compared individual differences in perseveration in *A. amazonica* across two housing conditions, as these differences could explain variability in the development and severity of stereotypic behaviors in captive animals. Parrots (n=13) were individually housed, from fledging, in the Baseline (BASE) condition, which included physical and foraging enrichment, and human socialization. They were trained on a standardized learning task (Hamilton Search Task), with phases allowing the measurement of two types of perseverative errors: recurrent (motor response; REC) and stuck-in-set (learning set; SIS). Following initial testing, the parrots were placed in the Unenriched condition (UNENR), i.e. standard laboratory housing (variable perches, ad lib food and water, and a wooden foraging cube) without additional enrichments or socialization. After 20 weeks in UNENR (a time period based on previous stereotypy research with *A. amazonica*), they were re-tested on the learning task. We found a large degree of between-subject variation in BASE perseverative tendencies in both REC (from 0.1 to 0.67, proportion of trials) and SIS (0.0 to 0.67) errors. The two types of error also responded differentially across conditions, with a significant test-retest increase in REC (t=1.96, P=0.0391 one tailed), but not SIS (t=-0.44, P=0.4 one tailed) errors. However, there was large between-individual variation in the magnitude of increase in both types of errors. The significant increase in REC errors post-UNENR supports theoretical arguments that recurrent perseveration may underpin stereotypies, as unenriched housing has been linked to stereotypic behavior in this species. We are now analyzing behavioral data to assess the relationship between baseline level of, and increase in, perseveration and the development of stereotypies during the UNENR condition.

Relationships between behaviour during temperament tests in prepubertal gilts with farrowing behaviour in two environments: a conventional crate or loose-housed (PigSAFE) pen

Ison, Sarah H.[1], Wood, Cynthia M.[2] and Baxter, Emma M.[1], [1]Scottish Agricultural College, Animal Behaviour & Welfare, Roslin Institute Building, Easter Bush, EH25 9RG Midlothian, United Kingdom, [2]Virginia Polytechnic Institute and State University, Department of Animal and Poultry Sciences, Blacksburg, 24061, USA; sarah.ison@sac.ac.uk

Sow welfare would be improved in loose-housed farrowing accommodation. Success in such systems depends on good maternal behaviour. Linking prepubertal gilt temperament with maternal behaviour would be of benefit when choosing gilts for loose-housed systems. Four groups of six (n=24) prepubertal gilts underwent three five minute human interaction and startle object tests, twice daily in a random order across three days. Duration of contact with the human or startle object was recorded. Approximately seven months later, 22 of these gilts went on to farrow in a conventional crate (C, n=11) or PigSAFE pen (PS, n=11), with straw provided daily and behaviour observed for eight hours from the birth of the first piglet (BFP). For the first human interaction and startle object tests, a mean was calculated to give contact duration for test one (CDT1) accounting for both tests. Behavioural variables during farrowing were analysed using GLMM with CDT1 and farrowing environment as fixed effects. Gilts in crates exhibited more piglet (C=11.01±1.99, PS=4.40±1.10% of time, P=0.005) and fixture (C=0.40±0.12, PS=0.017±0.091, P=0.02) focused behaviour, less straw /floor focused behaviour (C=1.03±0.30, PS=2.17±0.66, P=0.05) and less standing (C=3.13±1.15, PS=5.47±1.66, P=0.05). A significant negative relationship between CDT1 and time idle (P=0.02), lying lateral with udder fully exposed (P=0.02) and frequency of responses when piglets approach the head (P=0.04) and a positive relationship with standing (P=0.003) was found. Results indicate more contact with the human/startle object during testing is associated with higher activity during farrowing. It may be possible to select prepubertal gilts for less activity at farrowing, which would be of benefit particularly in loose-housed pens.

Response strategy of pigs in a delay discounting task is related to homovanillic acid excretion in urine but not to coping styles or aggression

Melotti, Luca[1,2], Toscano, Michael J.[1], Mendl, Michael[1] and Held, Suzanne[1], [1]University of Bristol, Animal Welfare and Behaviour Group, School of Veterinary Science, BS405DU Langford, United Kingdom, [2]University of Bern, Animal Welfare Group, Länggassstrasse 120, 3012 Bern, Switzerland; luca.melotti@vetsuisse.unibe.ch

We developed a novel delay discounting task to investigate impulsive choice in pigs, which could provide insight into their ability to cope with expectation and/or frustration, and reveal links between impulsivity, aggression, proactive/reactive coping styles, and dopamine and serotonin systems activity. Eight proactive and eight reactive pigs identified in pre-weaning manual restraint tests ('Backtest') were weaned at 26 d and mixed into four pens of four unfamiliar pigs, so that each pen had two proactive and two reactive pigs. Aggression was scored in the nine hours post-mixing. From age 1-3 months, pigs were individually tested in a delay discounting task, where they chose between a large and a small reward (four vs. one apple piece) during repeated trials. Initially, rewards were delivered with no delays and all pigs developed a preference for the large reward. Subsequently, the large reward was delivered after daily increasing delays (10 free-choice trials/day), and impulsivity was scored as the slope of the linear regression function between proportion of large reward choices and daily increasing delays, impulsive individuals being the ones discounting the value of the large reward quicker. Two strategies were adopted with some pigs switching their preference towards the small reward ('Switchers'), others persistently preferring the large reward until they stopped making choices ('Omitters'). Impulsivity was unaffected by such strategies and coping styles (2-way ANOVA, $F(1,12)=0.7$ and $F(1,12)=0.1$, respectively, n.s.), or by aggression at weaning ($r_s=0.14$, n.s.), yet basal urinary levels of homovanillic acid (dopamine metabolite), but not 5-HIAA (serotonin metabolite), were higher in Omitters ($F(1,10)=8.6$, $P=0.01$), and positively correlated with lever biting before choosing ($r_s=0.71$, $P=0.005$). Therefore, the two discounting strategies seemed to relate to the activity of the dopamine system yet not to coping styles, with Omitters prematurely terminating sequences of reward-directed behaviours, but not differing in impulsive choice, compared to Switchers.

Assessing emotional processes in dairy sheep of different temperaments: behavioural, cortisol and immune responses in dairy sheep experiencing positive and negative emotions on farm

Dimitrov, Ivan[1], Peeva, Jana[1], Sotirov, Lilian[2], Tsoneva, Vanya[3], Rassu, Salvatore P.G.[4], Mihaylova, Milena[1], Dimova, Nedka[1] and Vasilev, Vasil[1], [1]Agricultural Institute, Sheepbreeding, 6000 Stara Zagora, 6000 Stara Zagora, Bulgaria, [2]Tracian University, Veterinary Faculty, 6000 Stara Zagora, 6000 Stara Zagora, Bulgaria, [3]Tracian University, Medical Faculty, University Hospital, 6000 Stara Zagora, 6000 Stara Zagora, Bulgaria, [4]Sassary University, Dipartimento di Agraria, Via E. De Nicola, 9, 07100 Sassari, Sardegna, Italy; iv.dimitrov@dir.bg

The effect of temperament on positive (feeding, milking, behavioural contrasts) and stress-induced (human, shearing) biological reactions in dairy sheep, subjected to acute (shearing) and chronic (machine milking, human) stressors, and positive emotional stimulus (feeding, milking) on farm, was examined. Signs of pleasure (feeding, milking, affiliative behaviours) and fear (human, machine milking) have been used to assess the temperament in 235 dairy ewes, divergently selected for temperament (calm, nervous, dominant) for 21 years. Temperament was measured using a Complex Score (CS), reflecting the main behavioural traits of each animal during machine milking in milking parlour (verified every year). Traits were: (1) feeding activity towards neighbours; (2) feeding reaction towards foraging by hand; (3) reaction 'Positioning teatcups'; (4) persistency of taking place. Each trait was assessed by four degrees criteria. Observations were carried out on four consecutive days during morning milking. A Principal Component Analysis for behavioural traits produced three factors: Feeding, Milking and Social interaction, variation explained 0.439, 0.252 and 0.249. On the basis of CS, three temperaments ($P<0.001$), were established: Calm (C) – 75, Intermediate – 89 and Nervous (N) – 71 ewes. Temperaments were confirmed for consistency by different tests: fear (human, in all animals), learning (Pavlovian conditioning, in 40 sheep) and maternal care (selectivity alien/own lamb, all sheep). Endocrine responses of typical C (n=20) and N (n=20) ewes were measured after three treatments: 1) Machine milking; 2) Shearing; 3) Machine milking 24 h after shearing. Plasma cortisol levels were significantly higher in N ewes after machine milking (C – 8.1±1.4 nmol/l; N-23.7±6.1 nmol/l; $P<0.05$) and shearing (C – 19.9±5.2 nmol/l; N-38.5±6.1 nmol/l, $P<0.05$). Plasma lysozyme levels were higher after shearing for C ewes than for N ewes (C – 0.102±0.014 μg/ml; N – 0.065±0.005 μg/ml; $P<0.05$) but did not differ after the other treatments. This study shows that animals of different temperaments also expressed different biological responses to stressful events commonly encountered during modern sheep management.

Early tactile stimulation of piglets produces different subsequent behaviour responses correlated to body development

Oliveira, Daiana de[1], Keeling, Linda J.[2], Rehn, Therese[2], Zupan, Manja[2] and Paranhos Da Costa, Mateus J.R.[1], [1]São Paulo State University- Unesp, Animal Science department, Via de Acesso Prof. Paulo Donato Castellane, s/n, 14884900, Jaboticabal, SP, Brazil, [2]Swedish University of Agricultural Sciences-SLU, Department of Animal Environment and Health, Ullsväg 8C, P.O. Box 7038, Uppsala, Sweden; daiana_zoo@yahoo.com.br

The aim of this experiment was to study the role of early tactile stimulation of piglets on their reaction towards humans and growth rate. Female and castrated male piglets (66) from nine litters (Hampshire × Yorkshire) were subjected to 16 tactile stimulation sessions, starting at five days old. The tactile stimulation consisted of daily stroking on the back for two minutes. A handling resistance score was recorded (1=no resistant to 4=high resistant) after each session and a handling response index to reflect changes in resistance over time was developed by combining individual coefficient of variation (%) and the slope (in degrees) of the trend line through all sessions. From this, individuals in the first quartile were classified as sensitized and those in the last quartile as habituated. An open-field/human-approach test was conducted when piglets were 4 weeks old, with a familiar (FP) and an unfamiliar person (UP), measuring the number of squares crossed (activity), vocalizations and latency to approach FP and UP. Body weight was monitored at birth, 5, 9 and 12 weeks of age. Data were analyzed using procedure GLM and CORR. Of the piglets, 42.4% habituated to the handling whereas 31.8% were sensitized. Resistance score was positively correlated to weight gain and activity, meaning that high resistant piglets were less fearful in the arena ($r=0.26$ and $r=0.38$, respectively, $P<0.05$). Piglets discriminated between humans and reacted differently depending on how they had responded to the early tactile stimulation. Sensitized piglets were less active with a familiar person (FP=9.86±0.09) and more active with an unfamiliar (UP=11.62±0.04), whereas the pattern was the opposite for piglets that became habituated (FP=13.06±0.04 and UP=11.24±0.05; $P=0.01$). In summary, since reaction to early tactile stimulation was correlated to weight gain and appeared to be linked to the later human-animal relationship, a better understanding of early handling could improve both productivity and welfare in pigs.

How do lambs perceive regular human stroking: behaviour, cortisol and heart rate variability analyses

Coulon, Marjorie[1], Peyrat, Julie[1], Andanson, Stéphane[1], Ravel, Christine[1], Boissy, Alain[1], Nowak, Raymond[2] and Boivin, Xavier[1], [1]INRA, UMR1213 Herbivores, Site de Theix, 63122 Saint-Genès Champanelle, France, [2]INRA, UMR85 Physiologie de la Reproduction et des Comportements, Centre de Tours, 37380 Nouzilly, France; marjorie.coulon@clermont.inra.fr

Farm animal-human interactions vary in frequency between farmers, particularly with young, artificially fed animals. This study investigated the way lambs perceive regular human tactile contact using behavioural and physiological approaches. Twenty-four lambs were reared and bucket-fed in groups of four. All were stroked three times daily by a single person (SP). At 6 weeks of age, lambs were catheterised and tested two days later after 15 h of physical separation in a 1×1 m open-barred pen (OB-pen) within the rearing pen containing the rest of the group. Following a cross-over design once a week, they were tested without (control) or with the SP stroking the lambs for 9 min after 1 min of immobile presence. Blood samples were collected for cortisol determination (at -3 min, 1 min, 4 min, 10 min, 20 min). One week later, following the same test procedure, lambs were exposed to a known object (70cm-high traffic cone, previously exposed the day before the test). Another week later, a test was realised in which all lambs were separated in the OB-pen, then half of the lambs received stroking by the SP and half were exposed to the SP's immobile presence. Behavioural responses and heart rate were analysed before, during and after the human presence. Data were analysed with Proc-Mixed SAS procedure. Animals vocalised more after the human departure (4.3±0.9) than when tested alone (0.1±0.1) or with the object (2.5±0.8, $P<0.01$). They had more physical contact with the immobile SP than with the known object (35.7±4.6 vs. 19.7±3.0s; $P<0.01$). Across all tests, cortisol release did not vary ($P>0.4$). In contrast to passive human contact, stroking stimulated the parasympathetic system (logRMSSD-stroking=3.4±0.18 vs. passive=3.2±0.2, $P<0.05$) as well as after the human has left (logRMSSD-stroking=3.6±0.24 vs. passive=3.4±0.24, $P<0.01$). Gentle physical contact with the caregiver seems to be positively perceived by the lambs, possibly inducing a relaxed state and not simply habituation to such interaction.

Tying dairy calves for improving human-cows relationship seems not justified

Laffitte, Béatrice[1], Windschnurer, Ines[2], Schmied-Wagner, Claudia[2], Waiblinger, Susanne[2] and Boivin, Xavier[3], [1]Clinique vétérinaire des 110 Bêtes, 1030 rue du Gleysia, 64530 GER, France, [2]University of Veterinary Medicine Vienna, Instute Animal Husbandry & Animal Welfare, Veterinärplatz 1, 1210 VIENNA, Austria, [3]INRA, UMR1213 Herbivores, Centre de Clermont-Ferrand – Theix, 63122 Saint-Genes-Champanelle, France; laffitte.beatrice@orange.fr

The relationship between dairy farmers and their cows has a major impact on farmers' working conditions and animal welfare. Despite European legislation prohibiting tie housing for calves, many French farmers persisted in their practices. To understand this resistance, a questionnaire was sent out. 72 farmers responded. A large diversity in tethering practices (age, duration) was mentioned, with a third of the farmers reporting tethering their calves. Farmers who tied their calves claimed this practices resulted in calmer animals later in life. This notion was investigated further on 30 farms with cows born in the farm and differing tethering pratices in early age (regular tethering: n=12; individual boxes: n=13 and additionaly 5 farms with group housing: all bucket fed, same average herd size). Farmers were interviewed about their husbandry practices in early age and filled in a questionnaire on general and behavioural attitudes towards animals. Their behaviour towards cows was observed during milking. Avoidance distances of the cows towards an unfamiliar experimenter was recorded in the barn. Data were analysed through PCA (for attitude components) and variance/covariance analyses (for comparison of tethering practices). Variation between farms in cows' avoidance distances was mainly explained by rearing practices and the quantity of contact provided to calves. Cows' avoidance distances on the farms where calves were tied did not differ from those where they were reared in individual boxes (127±18 cm, P>0.1). However both had lower avoidance distances than those with their calves reared in groups (161±18 cm, P<0.01). In addition, farmers who reported to provide more contact to their calves had cows showing lower avoidance distances (P<0.01). Our results do not support the farmers' opinion about tying practices, but indicate the importance of positive contact to the animals. Our results can help the farmer to improve their animal docility without transgressing housing legislation.

Influences on avoidance and approach behaviour of dairy goats towards an unfamiliar person

Mersmann, Dorit, Schmied-Wagner, Claudia, Graml, Christine, Nordmann, Eva and Waiblinger, Susanne, University of Veterinary Medicine, Institute of Animal Husbandry and Animal Welfare, Veterinaerplatz 1, 1210 Wien, Austria; Dorit.Mersmann@vetmeduni.ac.at

Good human-animal relationship (HAR), characterized by animals with low fear of humans, can improve animal welfare and help avoiding handling problems. Larger herd sizes might lead to a reduced number of human-animal interactions and thus impaired HAR. We investigated goats' reactions towards humans using different tests and examined potential influencing factors on 38 large dairy goat farms (herd size 149±90 lactating dairy goats). In an approach test in the pen (group size 27 to 340) the number of goats approaching a stationary unfamiliar person within 3 m distance over 15 min (App_3m) was recorded. In the avoidance test the same testperson walked through the pen trying to touch as many individual goats as possible within 2 min (AvTouch=number of goats touched). Potentially influencing factors were assessed: Farmers' attitudes towards goats and towards interacting with goats via questionnaire, behaviour of stockpeople during milking was observed, intensity and type of human contact evaluated via interview (e.g. farmers' daily working time close to goats, experience in keeping goats in years) as well as animal, herd and, housing variables. Linear regression (stepwise forward) was performed after pre-selection of potentially influencing factors by Spearman correlation and Kruskal-Wallis Test. More goats could be touched (AvTouch) with lower percentage of negative interactions of the milkers (%NEG; β-coefficient: -0.55, P=0.000) and if stockpeople disliked more distinctly negative interactions with the goats (β: -0.31, P=0.022; model adj. R^2=0.42, P=0.000, n=36). As well more goats approached the human with lower %NEG (β: -0.52, P=0.001) and with lower number of visual interactions during milking (β: -0.46, P=0.002; model adj. R^2=0.37, P=0.000, n=35). In conclusion, these results suggest that reducing negative affective attitudes and negative interactions during milking can reduce dairy goats' fear of humans with potential beneficial effects on productivity and welfare. This appeared to be independent of herd size.

A positive mindset in the face of stress

Muns Vila, Ramon[1], Farish, Marianne[2], Rault, Jean-Loup[3] and Hemsworth, Paul[3], [1]Universitat Autònoma de Barcelona, Facultat de Veterinària, Bellaterra, 08193 Barcelona, Spain, [2]Scottish Agricultural College, Animal Behaviour and Welfare, West Mains Road, EH9 3JG, Edinburgh, United Kingdom, [3]University of Melbourne, Animal Welfare Science Centre, Parkville, 3010 Melbourne, VIC, Australia; raultj@unimelb.edu.au

Reducing stress in captive animals can be achieved by modifying the animal's environment through the identification and elimination of stressors, environmental enrichment or genetic selection that enhance stress resilience. Surprisingly there have been few attempts to examine strategies using positive classical conditioning to reduce stress, possibly through influencing emotional states. This study examined the effects of providing piglets with the opportunity to associate humans with feeding during the first day of life on their response to subsequent stressors imposed by humans. Forty litters from multiparous sows were randomly allocated to one of two treatments: Control (CON, minimal human interaction with day-old piglets) or Conditioning (POS, human talking and caressing day-old piglets during 6 suckling bouts). In each litter, 2 males and 2 females were randomly selected, tail docked at 2 days of age and caught and held for 10 s at 14 days of age before being submitted to a human approach and avoidance test. The behavioural response to tail docking and capture was assessed according to intensity (scale 0–4: no movement (0) to movement of high intensity (4)) and duration (0–3: no movement (0) to continuous movements (3)) and for the human test on occurrence (yes/no) of interaction for the approach test and withdrawal for the avoidance test. Statistical analyses were performed using GLIMMIX in SAS®. The POS piglets displayed a behavioural response to tail docking and capture that was less intense (P=0.09 and P=0.05, respectively) and of shorter duration (P=0.03 and P=0.03, respectively) than CON piglets. The POS piglets showed less avoidance to a human (P=0.08). The roles of classical conditioning, habituation and developmental changes in the effects of the POS treatment are yet unclear. Nonetheless, these preliminary results suggest a role for learning, with possible applications to enhance the ability to cope with stress.

Tickling makes handling a more pleasurable experience for laboratory rats

Cloutier, Sylvie, Panksepp, Jaak and Newberry, Ruth, Washington State University, Department of VCAPP, Center for the Study of Animal Well-being, P.O. Box 646520, Pullman WA, 99164-6520, USA; scloutie@vetmed.wsu.edu

Handling of laboratory rats can increase physiological and emotional stress, leading to a fearful relationship with humans. We hypothesised that the affective quality of handling techniques used with laboratory rats influences their fear of humans. In a series of experiments, we used ultrasonic vocalizations (USVs) production, and other behavioural measures to assess effects of various handling conditions on fear of humans in male Sprague-Dawley rats. Data were analysed as Mixed Model Anovas. When comparing rat responses to four handling treatments: (1) minimal handling once weekly (Control); (2) passive hand exposure (Passive, 2 min/day); (3) tickling, mimicking social play (2 min/day); and (4) restraint, intended to mimic pinning by a dominant rat (2 min/day), we found that Tickling and Restraint increased approach to a human hand ($P<0.05$) compared to Minimal handling. Tickling also increased production of 50-kHz USVs (a validated indicator of positive affect) in anticipation of handling ($P<0.0001$), and in the presence of a hand ($P<0.05$), compared to the Minimal and Passive treatments, respectively. Following an intra-peritoneal injection of saline, rats from Tickled and Passive treatments spent more time rearing near the experimenter than Minimally-handled and Restraint rats ($P<0.05$), indicating attraction to the experimenter. Furthermore, Tickling before, or before and after injection, induced a positive affective state, based on 50 kHz USVs production ($P<0.0001$). Tickling also eased handling during cage cleaning compared to Minimal handling ($P<0.0001$). When comparing preference for Tickling or Restraint handlers, rats spent similar time near both handlers, but nibbled their Tickler's hand (interpreted as play solicitation) more often than the hand of their Restraint handler ($P<0.05$). The results show that, although both the Passive and Restraint treatments had some value in habituating rats to humans relative to Minimal handling, Tickling was the most effective method for reducing rats' fear of humans and hence improving their emotional welfare.

Enhancing mouse welfare through non-aversive handling

Gouveia, Kelly and Hurst, Jane, University of Liverpool, Institute of Integrative Biology, Mammalian Behaviour & Evolution, chester high road, CH64 7TE, United Kingdom; kelly.gouveia@liverpool.ac.uk

Handling is a frequent cause of stress and variation in studies using laboratory rodents. Currently it is standard practice to pick up mice by the tail but recent evidence suggests that this procedure induces aversion towards the handler and high anxiety. By comparison use of a home cage tunnel leads to voluntary approach, low anxiety and acceptance of physical restraint. We need to understand whether handling tunnels need to be present in every home cage to be effective, and the importance of the duration of handling on taming animals. Our first study addressed the importance of tunnel familiarity on response to handling, while a second study investigated the impact of handling duration on response towards the handler. To assess response to handling we measured the willingness of mice to approach the handler. We compared handler interaction among mice picked up by the tail or tunnel, where groups differed in experience of a home cage tunnel. Use of a tunnel substantially enhanced approach compared to tail handling, even when animals were unfamiliar with tunnels ($F_{1,13}$=32.3, P<0.001). Tunnel familiarity initially increased handler interaction on day 1 ($F_{2,19}$=6.1, P<0.01), but regular tunnel handling removed this difference ($F_{2,19}$=0.05, P=0.96). The effect of handling duration on willingness to approach the handler was compared among mice handled by tunnel, tail or cupping on the open hand. Mice that were held for as little as 2s were as willing to interact as those handled for up to 60s (tunnel: $F_{3,24=}$0.4, P=0.78; non parametric ANOVAs for tail: $\chi^2_{3=}$4.3, P=0.23; cup: $\chi^2_{3=}$1.8, P=0.62). Thus, mouse welfare can be improved using non-aversive handling methods, even when handling is brief and without prior familiarity with handling tunnels.

In search of oestrus odours: do certain smells elicit penile erections in sexually naïve rats?

Nielsen, Birte L., Jerôme, Nathalie, Saint-Albin, Audrey, Rampin, Olivier and Maurin, Yves, INRA, UR1197 Neurobiologie de l'Olfaction et Modélisation en Imagerie, Bât 325, 78350 Jouy-en-Josas, France; birte.nielsen@jouy.inra.fr

Male rats display penile erections when exposed to faeces from mammalian females in oestrus, suggesting that specific odours may indicate female receptiveness across species. Identification of these compounds could be used in heat detection protocols for production species. However, it is unknown to what extent the sexual response is an odorous conditioning acquired during sexual encounters. We tested the behavioural response of 12 sexually naïve male Brown Norway rats (10 weeks old; no contact with females since weaning) when exposed to a potential oestrous odour (methylheptenone; a molecule found in higher concentrations during oestrus in rats, mares and vixens), and to rose odour (negative control) and oestrous rat faeces (positive control). Analyses of variance revealed a tendency for number of erections during the 30-min tests to be higher with both methylheptenone and the faeces compared to the rose (0.8, 0.9, and 0.2 erections (±0.2), respectively; P=0.068). Following sexual training (three sessions each with ejaculation and refractory period), the rats were tested again with the same odours and a novel negative control (herb odour). The rats showed higher levels of sexual arousal with both methylheptenone and the faeces compared to the negative controls (1.8, 2.4, 0.3 and 0.3 erections (±0.4), respectively; P<0.001) and to their naïve response to the same odours (P=0.082 and 0.007, respectively). The results indicate that sexually naïve male rats may have an innate affinity for certain odours associated with receptive females, and that this improves when the rats have gained sexual experience. The latter suggests that during sexual encounters an association is learned by the male rat between oestrous odours and the presumably rewarding experience of copulation. It remains to be seen if the innate response found in the present experiment reflects the odorous presence of a female, independent of her receptive state.

Using an appropriate reinforcement for triggering attention and enhancing learning performances: the example of horse training

Rochais, Céline[1], Henry, Séverine[1], Sankey, Carole[1], Górecka-Bruzda, Aleksandra[2] and Hausberger, Martine[1], [1]Université Rennes 1, UMR CNRS 6552, laboratoire Ethologie Animale et Humaine, Station Biologique, 35380 Paimpont, France, [2]Polish Academy of Sciences, Institute of Genetics and Animal Breeding, Jastrzebiec, Wólka-Kosowska, Poland; celine.rochais@univ-rennes1.fr

In horses, the use of a positive reinforcement as a food reward has been shown to enhance learning performances and to promote a positive relation to the trainer. Here, we investigated whether grooming at the whither, known to induce a decrease in the groomee's heart rate, could be used as a primary reinforcement, by comparing it to a food reward. Twenty Konik horses (9 females, 11 males), aged 1-2 years, with no previous learning experience and no contact with humans, except for feeding, were trained to remain immobile in response to a vocal command. Training was performed 5 minutes per day for 6 days. Horses were allocated to one of two training groups: in the food-reward group (FR: n=10), the trainer hand-gave a piece of carrot to the horse after it responded correctly, while in the grooming-group (GR: n=10), the trainer scratched the horse's wither. Duration of immobility and trainer-directed behaviours were continuously recorded. The results indicate that: (1) using food rewards facilitated learning (duration of immobility: $P<0.05$), whereas grooming rewards led to lower performances; (2) the type of reward has a clear effect on the horse's attentional state during training. While no difference between both groups were scored in the first days of training, on the last day the FR group spent more time monitoring (i.e. head rotations towards the trainer), more time gazing at and expressed more investigative behaviours towards the trainer (e.g. sniffing) than the GR group (MW, $P<0.05$ in all cases). In conclusion, these results suggest that the use of a food reward is associated with an increase of motivation, leading to a better efficiency in promoting learning that may well have been mediated by attentional factors, while alternative such as scratching (usually use in training techniques) appears less efficient.

Pigs' use of human social cues and attentive states in an object choice task

Nawroth, Christian[1], Ebersbach, Mirjam[2] and Von Borell, Eberhard[1], [1]Martin-Luther-University Halle-Wittenberg, Department of Animal Husbandry & Ecology, Theodor-Lieser-Str. 11, 06120 Halle, Germany, [2]University of Kassel, Department of Developmental Psychology, Holländische Str. 36-38, 34127 Kassel, Germany; christian.nawroth@landw.uni-halle.de

The use of human social cues and attentive states by eleven domestic, socially housed pigs (16 wks of age, 6 females) was investigated. In Experiment 1, subjects had to choose between two buckets of which one was baited, indicated by four different social cues given by the experimenter (tap, point, body orientation, head + gaze). Only three pigs finished the first session without a side bias and were therefore tested in eight further sessions. They performed as a group significantly above chance with the tapping and pointing cue ($P<0.05$) but not with body and gaze cues. Analyzed individually, two pigs developed a side bias for body and gaze cues during the second half of the experiment. Analyzing only the first half, one subject used the tapping and pointing cue as well as body orientation to locate the correct bucket (8 out of 9 trials correct; $P<0.05$ for every condition). In Experiment 2, subjects had to choose between two unfamiliar experimenters with only one showing attention, looking straight at the subject, during different conditions (body turned, head turned, body turned – head front). Three subjects were excluded from the analysis due to a side bias or not making any choices. The remaining subjects (n=8) performed as a group significantly above chance in every condition (body turned: 0.59±0.27, head turned: 0.59±0.19, body turned-head front: 0.59±0.19). As there were no performance differences between conditions, individual performance was pooled, revealing a significant preference of pigs to approach the experimenter who was looking at them ($P<0.05$). The results suggest that pigs might be able to use social cues and attentive states of humans to obtain information. The role of side biases, learning and local enhancement will be discussed.

Can dogs relax: work-related cortisol levels vary in dogs during animal-assisted interventions

Glenk, Lisa M.[1,2], Stetina, Birgit U.[3], Kothgassner, Oswald D.[4], Kepplinger, Berthold[2] and Baran, Halina[2], [1]University of Veterinary Medicine, Veterinärplatz 1, 1210 Wien, Austria, [2]Karl Landsteiner Research Institute for Pain Treatment and Neurorehabilitation, Neuropsychiatric Hospital Mauer, 3362 Amstetten/Mauer, Austria, [3]Department of Psychology, Vienna Webster University, Berchtoldgasse 1, 1220 Vienna, Austria, [4]Institute of Clinical, Biological and Differential Psychology, Liebiggasse 5, 1010 Wien, Austria; lisa.molecular@gmail.com

Positive effects of human-animal contact on human health have contributed to the wide distribution of animal-assisted interventions (AAIs). While considerable effort has been devoted to the research on human welfare associated with AAIs, potential effects on therapeutic animals have received little attention. Therapeutic dogs are required to cope with stressful conditions, deal with unfamiliar people and strange situations. The aim of this study was to determine baseline and work-related levels of cortisol, a glucocorticoid hormone that is known to vary with physiological arousal, in therapeutic dogs. Fourteen certified therapy dogs aged 5.3±3.9 years (Mn ± SD), 7 dogs in AAI program 1 (P1; dogs on-lead during work) and 7 dogs in AAI program 2 (P2; dogs off-lead during work), participated in the study. Pre-post (on 2 AAI working days) and baseline (at home) salivary samples were collected and analyzed with enzyme immunoassay. Statistics included Friedman two-way ANOVA in evaluating home levels and repeated measures ANOVA for working sessions with intervention type as between-group factor and time as repeated factor. There was no difference between cortisol home baseline in P1 (5.7±2.4 ng/ml) and P2 dogs (5.0±2.0 ng/ml). However, significant differences between P1 und P2 were found on days with therapeutic work (P=0.010). In group P2, the decrease in cortisol from session baseline to working levels was -2.45 ng/ml on working day 1 and -2.34 ng/ml on working day 2. Group P1 showed a decrease in cortisol of -0.59 ng/ml on working day 1 and an increase of +0.28 ng/ml on working day 2. Salivary cortisol levels in therapeutic dogs performing AAI P1 and P2 vary significantly. These insights suggest that dogs which are off the lead during intervention show decreases in working cortisol and hence, might be more relaxed.

The relationship between human and dog behaviour in animal shelters

Haynes, Sally J.[1]*, Coleman, Grahame J.*[2] *and Hemsworth, Paul H.*[1,3]*,* [1]*The University of Melbourne, Animal Welfare Science Centre, Melbourne School of Land and Environment, Parkville, Victoria, 3010, Australia,* [2]*Monash University, School of Psychology, Psychiatry and Psychological Medicine, Clayton, Victoria, 3800, Australia,* [3]*Department of Primary Industries, Animal Welfare Science Centre, 600 Sneydes Road, Werribee, Victoria, 3030, Australia; sjhaynes@me.com*

The human-dog relationship at four animal shelters was studied by observing the behaviour of 617 individually-housed dogs and 29 handlers during pen hosing, using one-zero sampling. The occurrence or not of the individual 'positive handler behaviours' of bend, crouch, extend hand, talk, pat/stroke and pick up the dog and the individual 'negative handler behaviours' of shout, drop object, bang gate, push dog and hose dog were recorded. The 'fearful dog behaviours' of crouch, head oriented away, tail low or tucked and tail still were recorded using one-zero sampling. Minimum distance between dog and handler and distance between the two on entry and exit were also recorded. Stepwise forward regression analyses were used to examine the relationships between human and dog behaviour. The model that best predicted minimum distance between dog and handler in the pen included the numbers of individual positive handler behaviours, individual fearful dog behaviours and individual negative handler behaviours (adjusted R^2=0.226, $F_{1,604}$=60.1, P<0.001). The direction of this relationship indicates that the numbers of individual positive and negative handler behaviours were associated with reduced minimum distance between the dog and handler, whilst the number of individual fearful dog behaviours was associated with increased minimum distance between the dog and handler. Furthermore, neither the number of individual positive nor negative human behaviours was correlated with distance between the dog and handler on the handler's entry to the pen, which suggests that handlers were not modifying their behaviour on the basis of initial dog behaviour. Very similar correlations were observed for data collected during other cleaning activities such as solid waste removal. These relationships, although not conclusive evidence of causal relationships, highlight the need for further research to examine the effects of human behaviour on dog behaviour and the opportunities to apply this knowledge to improve dog behaviour and welfare.

Does origin litter affect social rank of piglets after weaning?

Fels, Michaela and Hartung, Jörg, University of Veterinary Medicine Hannover, Foundation, Institute for Animal Hygiene, Animal Welfare and Farm Animal Behaviour, Bünteweg 17p, 30559 Hannover, Germany; michaela.fels@tiho-hannover.de

It is common practice in pig production to mix unfamiliar piglets of uniform weight after weaning often leading to severe fighting aimed at establishing a social hierarchy. The aim of this study was to investigate the influence of origin litter on agonistic behaviour and the social rank of piglets obtained within 3 days after mixing. 120 piglets from 20 litters were studied in groups of 12 with 6 piglets from 2 litters each in 5 rearing rounds. The mean initial weight was 9.9 kg with an average age of 35 days. The animals were kept on fully slatted floor with 0.3 m^2 per animal. Piglets had *ad libitum* access to dry food and water; animal-feeding place ratio was 1.5:1. Groups were formed as homogeneous weight groups with equal sex ratio. The number and outcome of all agonistic interactions were analyzed during 72 hours after mixing. Individual rank indices (RI) were calculated based on the number of wins (S) and defeats (N), the number of partners against each piglet has won (P_S) or lost (P_N) and the group size (n): RI = $(SxP_S)-(NxP_N) / (S+N)x(n-1)$. RI ranked from -1 (absolutely subdominant) to +1 (completely dominant). ANOVA analysis followed by posthoc test (SNK) was conducted. Piglets originated from one litter obtained higher rank indices than piglets originated from the other litter within the same group ($P<0.05$). The highest ranking piglet in a group showed more attacks than lower ranking animals (12.7 vs. 3.4, $P<0.05$) fighting particularly against non-littermates. We conclude that piglets obtain a litter-associated dominance status when mixing an equal number of piglets from two different litters after weaning. We assume that the piglets of one litter cooperate after mixing, and thus are able to evaluate their own fighting success based on the outcomes of fights between their littermates and non-littermates.

Behaviour of finishing pigs divergently selected for social genetic effects in barren and straw-enriched pens

Camerlink, Irene[1,2], Bijma, Piter[1], Ursinus, Winanda W.[2,3] and Bolhuis, J. Elizabeth[2], [1]Wageningen University, Animal Breeding and Genomics Centre, P.O. Box 338, 6700 AH Wageningen, the Netherlands, [2]Wageningen University, Adaptation Physiology Group, P.O. Box 338, 6700 AH Wageningen, the Netherlands, [3]Wageningen UR Livestock Research, Animal Behaviour and Welfare, P.O. Box 65, 8200 AH Lelystad, the Netherlands; irene.camerlink@wur.nl

Group-housed pigs may influence each other's behaviour and growth through social interactions. The genetic effect of a pig on the growth of its group mates during the finishing period can be estimated, and is referred to as its Social Breeding Value (SBV). At present, behavioural consequences of selection for SBV are unknown. We, therefore, studied the behaviour of pigs selected for either high or low SBV in different housing conditions. In five batches, a total of 480 finishing pigs with intact tails were studied in a 2×2 arrangement with SBV and housing (barren vs. straw-enriched) as treatments. Pigs were housed in groups of six in 7 m^2 pens (80 pens). Pigs were observed at 12, 16 and 21 weeks of age by 2-min instantaneous scan sampling for 6 h during daytime. Behaviours were averaged per pen and over observations and analysed using GLMs with SBV, housing, their interaction, and batch as effects. High SBV pigs showed 15% more comfort behaviour than low SBV pigs ($P<0.05$). Pigs selected for low SBV spent 14% more time on ear biting ($P<0.05$) and spent 35% more time on chewing a jute sack or metal chain ($P<0.01$). Low SBV pigs indeed used up more jute sacks ($P<0.001$), but also had higher tail damage scores ($P<0.05$). This may indicate that low SBV pigs have a stronger tendency to perform oral manipulation. Straw-bedding strongly reduced time spent on ear biting, tail biting and other oral manipulative behaviours, up to a 64% reduction of tail biting behaviour (all $P<0.001$). Straw-bedding increased exploration with 26% ($P<0.001$) and comfort behaviour with 25% ($P<0.001$). These results suggest that both selection for high SBV and enriched housing may enhance pig welfare through positive effects on (social) behaviours.

The social structure of farmed female capybaras

Nogueira-Filho, Sérgio L., Lopes, Pauliene, Nobrega, Djalma, Santos, Ednei and Nogueira, Selene, Universidade Estadual de Santa Cruz, Laboratório de Etologia Aplicada, rod Ilheus Itabuna km 16, Ilhéus, Bahia, 45662-900, Brazil; isaelailheus@gmail.com

Capybaras (*Hydrochoerus hydrochaeris*) live in stable social units composed of adult males and females with their young. It has been confirmed that a linear dominance hierarchy characterizes interactions among males. There is, however, no information on dominance relationships among females. Therefore, this study described the social structure of farmed female capybaras living in five groups, composed of one male and five to 13 females, kept in paddocks from 1000 to 4,500 m^2. The groups were caught in the wild between 2004 and 2008 and kept thereafter in enclosures with earth underfoot, low bushes, one water tank (6.0×5.0×1.0 m), one water trough (0.6×0.3 m), and one feeder (1.0×0.3 m). All occurrences of aggressive behavioral patterns (chasing, threatening, pushing, and attempts to bite) were recorded during 2 h observation sessions at feeding time, totaling 40 hours of data collection per group. To test for linearity of hierarchy, Landau's corrected linearity index (h') was calculated using SOCPROG (2.4) software. Additionally, the live weights of the subjects were correlated with their rank positions by the Spearman rank coefficient (r_S). Dominance relationships in capybara females fit a linear hierarchy since h' ranged from 0.73 (P=0.0001), in the largest group, to 0.97 (P=0.021) in the smallest one. The dominant female tends to be heavier (r_s ranged from 0.79 to 0.82, $p<0.02$). Males did not influence female social structure. High-ranking capybara females may obtain access to the best foods, allowing them to invest more energy in reproduction and produce more offspring. They can also monopolize mating partners and secure exclusive mating. The obtained results may explain the great variation in reproductive parameters recorded among farmed capybaras. Therefore, management practices must be established to improve health and welfare of low-ranking females and to guarantee these females access to resources.

Pair-bonding and companion recognition in domestic donkeys (*Equus asinus*)

Murray, Leigh M.A.[1], D'Eath, Richard B.[1] and Byrne, Katharine [2], [1]Scottish Agricultural College, Animal Behaviour and Welfare, Animal and Veterinary Sciences Research Group, West Mains Road, Edinburgh, EH9 3JG, United Kingdom, [2]University of Edinburgh, College of Medicine and Veterinary Medicine, 49 Little France Crescent, Edinburgh, Midlothian EH16 4SB, United Kingdom; leigh_murray75@hotmail.co.uk

Pair and social bonding has been documented in various taxa, where they are usually driven by kinship or sexual motivation. However, pair-bonding between unrelated individuals without sexual motivation is not well documented. The aims of this study were (1) to investigate the existence of pair-bonding in domestic donkeys; and (2) to determine whether members of a dyad could recognise their companion during a Y-maze recognition test. Subjects were 55 unrelated donkeys (38 geldings, 17 mares) from seven mixed or same sex groups, comprising 4-14 individuals. Spatial proximity was observed three times a day over a 22 day period. During these observations, donkeys were free-ranging in their home fields ranging in size from 0.81-2.43 ha, with access to indoor and/or outdoor field shelters, food and water. Using a simulation approach based on observed data to generate randomised matrices, the statistical significance of social relationships was estimated. Results show 42 of 53 donkeys (79.2%) were involved in significant ($P<0.05$) non-random relationships, most of which (71.7%) were reciprocal pair relationships. Results from spatial data revealed 24 donkeys had shown significant reciprocal preferences. These were then tested in a Y-maze recognition test in which they were presented with a choice of 1) their companion and a familiar donkey (same group) or 2) their companion and an unfamiliar donkey (different group). Both donkeys in a companion pair were used as test subjects on different occasions in the Y-maze test, with each donkey completing six tests in total. Donkeys' spatial location in the Y-maze demonstrated a preference for their companion whether the other donkey was familiar (Wilcoxon signed rank test $W=239$, $P=0.002$) or unfamiliar ($W=222$, $P=0.041$). This supports anecdotal evidence that donkeys form pair-bonds, and shows that reciprocal social preference and recognition are the basis of these.

Do you smell what I smell: using olfactory cues to modulate the social behavior of laboratory rabbits

Valuska, Annie and Mench, Joy, University of California, Animal Behavior Graduate Group and Center for Animal Welfare, One Shields Avenue, Davis, CA 95616, USA; jamench@ucdavis.edu

While social housing of laboratory rabbits is considered desirable for welfare, mixing of unfamiliar does is difficult because of injurious aggression. In wild rabbits, conspecific scent marking plays a role in regulating social behavior. Because bucks scent mark more than does, we evaluated the possibility that buck urine scent could decrease aggression among unfamiliar does during initial introduction. Does were individually housed and tested in a series of unique pairings for 1 hour. In the first experiment, either a dry cotton pad (CONTROL) or one containing 2 mL of buck urine (URINE) was applied to the foreheads of pairs (n=22 pairs total) of unfamiliar does. The does were then introduced to one another in an apparatus comprising 2 pens separated by a PVC-pipe barrier that allowed olfactory and tactile contact, and observed continuously using video recording. Data were analyzed using a general linear model. URINE pairs were significantly more affiliative (mean=35.9 vs. 20.3 interactions; P=0.02) and less aggressive (0.24 vs. 0.61 interactions; P=0.01) than CONTROL. To determine whether this effect was due specifically to male odor rather than to the does having a common scent, a second series of experiments was conducted in which unfamiliar doe pairs were tested after application of urine from the same buck, urine from different bucks, or urine from does (n=11 pairs per treatment). Pairs marked with buck urine were less aggressive (median=0) than doe urine pairs (median=2; Kruskall-Wallis test, P=0.04) during initial introduction. In contrast, there were no differences in affiliation or aggression between pairs marked with urine from the same or different bucks. Application of buck urine could be an effective tool for reducing aggression between does during introductions, although its long-term efficacy remains to be evaluated.

Effects of group stability on aggression, stress and injuries in breeding rabbits

Roth, Beatrice A.[1]*, Andrist, Claude A.*[1]*, Würbel, Hanno*[2] *and Bigler, Lotti M.*[1]*,* [1]*Swiss Federal Veterinary Office, Centre for Proper Housing: Poultry and Rabbits, Burgerweg 22, 3052 Zollikofen, Switzerland,* [2]*University of Bern, Division of Animal Welfare, Länggassstrasse 120, 3012 Bern, Switzerland; beatrice.roth@bvet.admin.ch*

Regrouping female breeding rabbits in group-housing systems is common management practice, which may induce agonistic interactions resulting in stress and injuries. On farms using artificial insemination, does are kept singly for 12 days around parturition to avoid pseudogravity and fighting for nests. The integration of new group members occurs after this isolation phase. We studied whether keeping groups stable would reduce aggression, stress and injuries after reunion. Does were kept in 24 pens containing 7-8 rabbits each. In 12 pens, group composition before and after the 12 days isolation period remained the same (S-groups). In the other 12 pens three randomly chosen does were replaced after the isolation phase by three unfamiliar does (M-groups). S-groups were together for one reproduction cycle. Once before and three times after regrouping, agonistic interactions (continuously, 2×4 h), stress levels (corticosteroids, collection: 12 days before, 36h/84h/132h after) and injuries (score: no injuries=0, maximum=3) were recorded and analysed using mixed-effects-models. After regrouping, the frequency of agonistic interactions were increased in both groups, but tended to increase more in M-groups (e.g. boxing, M: before 0.09±0.09 and 3rd day: 2.27±1.07; S: before 0.83±0.44 and 3rd day: 2.17±1.07/pen/8h, treatmentXday: P=0.086). There were no treatment effects on stress levels (P=0.199) and injury scores (P=0.589). Injury scores were highest on the second day after regrouping and decreased in both groups (M: 2nd day: 0.45±0.08, 4th day: 0.26±0.06, 6th day: 0.24±0.06; S: 2nd day: 0.33±0.07, 4th day: 0.28±0.06, 6th day: 0.21±0.06/animal/day, P=0.015). This suggests that keeping groups stable for one reproduction cycle had little effect on aggression, stress and injuries. Although agonistic interactions tended to increase more in M-groups, aggression and injury scores were elevated in both groups. One reproduction cycle might be too short to maintain a stable hierarchy, which after 12 days of separation may need to be re-established.

Characteristics of social relationships in a dynamic dairy cow herd

Gutmann, Anke K.[1], Špinka, Marek[2] and Winckler, Christoph[1], [1]University of Natural Resources and Life Sciences, BOKU, Department of Sustainable Agricultural Systems, Division of Livestock Sciences, Gregor-Mendel-Strasse 33, 1180 Vienna, Austria, [2]Institute of Animal Science, Department of Ethology, Přátelství 815, 104 00 Praha Uhříněves, Czech Republic; anke.gutmann@boku.ac.at

Cattle are a highly social species, naturally living in stable groups with long-lasting relationships. Only little is known about characteristics and role of relationships in unstable groups, as usually found in dairy farms. As a first step, the aim of the present study was to investigate how relationships between cows can be characterised. Data were collected in a herd of 48-54 cows in early lactation on a 230-cow farm. Over an 8-weeks-period, groups of focal cows (4-6 each, n=18) were continuously observed for five days using video recordings. Behavioural data comprised social interactions (SI), and lying and feeding neighbours. Data were converted into directed relative values per dyad, i.e. for example how many times a focal cow (F) was displaced by a partner (P) relative to the total number of encounters of this pair, and how often F displaced P. A principal component analysis was applied on a total of 17 variables to reveal related and unrelated behavioural aspects of relationships. Preliminary data analysis of the first observation day of eight cows (n(dyads)=187) revealed four components explaining together 60% of variance. Component 1 represents an 'avoidance'-axis defined by submissive behaviours and explained 22% of variance, Component 2 explained 17% and reflected 'P's tolerance' (consequence of SI for F ranging from neutral to displacement/withdrawal), Dimension 3 explained 11% and ranged from 'chose as neighbour' to 'chosen by neighbour' ('decision maker'), and Component 4 explained 10% and reflected 'F's tolerance'. These preliminary results confirm assumptions about characteristics of social relationships in cattle defined by avoidance versus tolerance and reduced aggression. Analysis of a larger data set will be presented showing how relationships in a dynamic dairy herd are distributed and whether they are influenced by factors such as age, familiarity, or time spent in the herd.

Influence of social hierarchy on the use of shade by dairy cows

Pinheiro Machado Filho, L. Carlos, Pellizzoni, Camila, Weickert, Lizzy, Avila, Taís S. and Hötzel, Maria J., LETA – Lab. de Etologia Aplicada e B-E Animal, Universidade Federal de Santa Catarina, Zootecnia e Des. Rural, Rod. Admar Gonzaga, 1346, 88.034-001 Florianopolis, SC, Brazil; pinheiro@cca.ufsc.br

When resources are limited (e.g. shade, water, etc.), competition may interfere in their welfare. The quality of the resource in dispute may also affect the intensity of the competition. The aim of this study was to evaluate the influence of social hierarchy on the use of different shading material and space per animal, and its consequences on behaviour, respiration rate and rectal temperature in four groups of cows on pasture. In a Latin square design, four groups of 4 or 5 animals were allocated to all following treatments: 6 m^2 screen (S6), 18 m^2 screen (S18), 6 m^2 tile (T6) and 18 m^2 tile (T18), in periods of 3 days, averaging 26 °C. Screen was 70% black plastic cloth, and tiles were made by clay. Behavioural data were collected in days 2 and 3 through direct observations between 10:00 and 16:00. Respiration rate was collected at 10:00, 13:00 and 15:00, and rectal temperature was collected at milking times (6:00 and 16:00). For each group we calculated the social rank using a sociometric matrix. Differences between treatments were determined using analysis of variance. Subordinate cows used shade less time on screen but not on tile treatments (subordinate, intermediate and dominant cows spent, respectively, in S6: 50, 69, 89; in S18: 45, 82, 87; in T6: 75, 78, 84; in T18: 81, 84, 87% of the observed time; $P<0.05$). However, there was a higher incidence of aggressive interactions in treatments with little shade amounts (S6=12±1, S18=5±1, T6=13±1, T18=7±1 events/observation; $P<0.05$). The respiration rate was greater for S6 compared with the other three treatments (88±2 versus 75±2 mov/min; $P<0.01$) but we didn't observe a treatment effect on rectal temperature. This study confirmed the influence of social hierarchy on the use of shade. However the influence of social hierarchy seems to be dependent on the quality of the shade.

Housing effects on social licking in dairy heifers

Tresoldi, Grazyne[1,2], Weary, Daniel M.[2], Pinheiro Machado Filho, Luiz Carlos[1] and Von Keyserlingk, Marina A.G.[2], [1]LETA – Lab. de Etologia Aplicada e Bem-estar animal, Dpto de Zootecnia e Desenv. Rural, Universidade Federal de Santa Catarina, Rod. Admar Gonzaga, 1346, Itacorubi, Florianópolis, 88034-001, Brazil, [2]Animal Welfare Program, Faculty of Land and Food Systems, University of British Columbia, 2357 Main Mall, Vancouver, V6T 1Z4, Canada; tresoldi.g@gmail.com

The type of housing has been shown to dramatically affect social behaviors, such as competition, in farm animals. Little work, however, has addressed the impact of housing on social licking, an affilliative behavior normally associated with good welfare in farm animals. The aim of this study was to describe social licking in dairy heifers housed indoors in a freestall barn and on pasture. We used 6 groups of 8 pregnant Holstein heifers each, tested in both conditions using a crossover design. Details of each social licking event (executor-receptor involved, engaged behavior, body part licked, start-termination and duration) and agonistic (executor-receptor involved) bout were collected through direct observation during 2 periods of 6 hours each in both treatments. Differences between the two housing environments were determined using Chi-square and t-tests. Irrespective of housing treatment 70% of the social licking bouts started without solicitation. In the freestall barn, 87% of these licking bouts were terminated without conflict, versus 71% of bouts on pasture (P=0.0036). There was no housing effect on duration of licking events (averaging 34.6±7.1 s per bout), body part licked (46% towards the head) and engaged behavior (62% of lickings occurred while animals were feeding/grazing). The total number of social licking and agonistic events was 4 times higher when heifers were housed in the freestall barn compared to pasture (546±43 vs. 128±7 events per group, P=0.015), but number of social licking bouts as a ratio of the total amount of events was the same in both treatments (Freestall barn: 0.11±0.02 versus Pasture: 0.07±0.01; P=0.27). In summary, heifers engaged in more social interactions when housed indoors but this increase included both agonistic and social licking events. The increased interactions are likely due to increased proximity of animals when housed indoors.

Temporal separation and subsequent reintegration of individual goats: assessment of welfare effects

Patt, Antonia[1,2], Gygax, Lorenz[2], Hillmann, Edna[1], Palme, Rupert[3], Wechsler, Beat[2] and Keil, Nina M.[2], [1]ETH Zurich, Institute of Agricultural Sciences, Universitätstrasse 2, 8092 Zurich, Switzerland, [2]Federal Veterinary Office, Centre for Proper Housing of Ruminants and Pigs, Agroscope Reckenholz-Tänikon Research Station ART, Tänikon, 8356 Ettenhausen, Switzerland, [3]University of Veterinary Medicine, Department of Biomedical Sciences/Biochemistry, Veterinärplatz 1, 1210 Vienna, Austria; antonia.patt@art.admin.ch

Goats are very likely stressed when separated from their group as well as during subsequent reintegration. Effects of two treatments of separation and return were tested by separating 12 female goats of different ranks individually for two days from their groups (4 in total). In the 'contact'-treatment, the separated goat had tactile, visual and acoustic contact to its group through bars, contrary to only acoustic contact in the 'no-contact'-treatment. Each goat experienced the two treatments in balanced order. Effects of separation and return were assessed by measuring social interactions, lying and feeding behaviour, the incidence of injuries, and the concentration of cortisol metabolites in the separated/returned goats' faeces. Data was analysed using generalised linear mixed-effects models with 'day' of the observation period (control: day -2 and -1, separation: day 0-1, return: day 2-4), 'treatment' and 'rank' and their two- and three-way interactions as explanatory variables. Compared to control days, time spent feeding was decreased (day 0: -0.75 hours/day, confidence interval [-1.31; -0.19], P=0.004) and concentration of cortisol metabolites was increased (day 1: +83%, [+37; +144], P<0.001) during separation irrespective of treatment. Additionally, goats in the 'no-contact'-treatment lay less during separation than goats in the 'contact'-treatment (P<0.001). On the first day of return (day 2), the probability that individual goats directed sniffing (OR=5.20, [2.08; 13.10], P<0.001) and agonistic behaviour (OR=1.94, [1.09; 3.42], P=0.03) towards residents increased compared to control days. Furthermore, cortisol metabolites were elevated on day 2 (+54%, [+15; +105]; P<0.001) and tended to be lower in the 'contact'-treatment (-19%, [-36; +2], P=0.075). Rank didn't significantly influence lying and feeding nor cortisol metabolites concentration. We conclude that both, separation of individual goats from their group and, to a smaller extent, the return to their group adversely affect their welfare. Separation allowing tactile, visual and acoustic contact mitigates such effects.

Measures of acceleration can estimate the effects of dehorning and weaning on the duration of locomotor play in calves

Rushen, Jeffrey and De Passillé, Anne M., Agriculture and Agri-Food Canada, 6947 Highway 7, Agassiz, BC V2P 7P9, Canada; jeff.rushen@agr.gc.ca

Play behaviour is a behavioural indicator of positive emotions in young animals, but in calves, play occurs rarely and is time consuming to record. Research shows that dehorning and weaning off milk reduce spontaneous play running that calves show in their home pens. We examined whether automatic measures of acceleration during brief exposure to a large enclosure could estimate the duration of the play running that occurred and detect the effects of dehorning and weaning. Holstein calves were placed in an arena for 10 mins and the duration of running was recorded from video. Total acceleration was recorded from accelerometers attached to one leg. In Exp. 1, 24 calves were either dehorned on d5 or served as controls, and were tested the days before and after dehorning. Dehorned calves had a lower duration of running (median ± quartile deviation = 22.4±20.1 s vs. 31.8±21.4 s; $P<0.01$; Mann-Whitney test) and a smaller total acceleration (24,517±1,542 G vs. 25,115±672 G; $P<0.01$) than controls. Before and after dehorning, duration of running was correlated with total acceleration (both $r>0.77$; $P<0.001$). In Exp. 2, 12 calves were tested during the weeks before and after weaning at 3 months of age. Before and after weaning, the duration of running was positively correlated with total acceleration (both $r>0.85$; $P<0.001$). After weaning, there was a decrease in both the duration of running and the total acceleration (Mann-Whitney; $P<0.01$), and the change in running duration following weaning was correlated with the change in total acceleration ($r=0.94$; $P<0.01$). The amount of running the calves did in the arena was reduced following both dehorning and weaning, and measures of acceleration could be used to automatically estimate the duration of running that occurred, suggesting this was a time-efficient way of detecting the calves' motivation to play run.

Open-field behaviour in young rabbits – how much do the lines tell us?

Buijs, Stephanie, Maertens, Luc and Tuyttens, Frank A.M., Institute for Agricultural and Fisheries Research, Animal Sciences Unit, Scheldeweg 68, 9090 Melle, Belgium; stephanie.buijs@ilvo.vlaanderen.be

The number of lines crossed is commonly scored in open-field tests. If this measure explains the majority of the behavioural variance, analysis may be simplified by scoring crossings only. To see if this is the case for young rabbits, 21 individuals were subjected to open-field testing (1×1 m open-field divided into 0.2×0.2 m squares, age: 16, 17 and 18 days), and novel-object testing (a multicoloured stick lowered into the open-field, age: 25 days). In the 5 minute open-field test we scored the number of crossings, and the time spent hopping (movement involving synchronous hind leg motion), walking (movement involving alternating hind leg motion), rearing or jumping, stationary whilst scanning the environment, and stationary without scanning the environment (including grooming). In the novel-object test we scored approach latency. Stationary whilst scanning (52%± 8SD) and hopping (23% ±6SD) were most common in the open-field. Principal component analysis summarized related behaviours into three independent components with eigenvalues >1. Eigenvectors >|0.5| were used to interpret these components. Component 1 explained 38% of the variance, reflecting quick/ normal locomotion (positive eigenvectors for line crossings and hopping). Component 2 explained 26% of the variance, reflecting attention to the environment (positive eigenvector for stationary without scanning and negative eigenvector for stationary whilst scanning). Component 3 explained 20% of the variance and reflected tentative locomotion (positive eigenvector for walking, which is slower than hopping). Component 2 increased with test-day (repeated measures ANOVA, $F_{2,40}$=5, P=0.01), suggesting that attention to the environment declined with decreasing novelty. Animals that were less attentive during the first open-field test (higher component 2 values) tended to approach the novel-object quicker (r=-0.41, P=0.08). No such effects were found for component 1 and 3. Together, this suggest that attention to the environment during the open-field test reflects rabbits' response to novelty, thus providing important information additional to line crossings.

Three kinds of agreement: bias, association and exact matching

D'Eath, Richard B., SAC, Animal Behaviour & Welfare, Animal & Veterinary Sciences Research Group, West Mains Road, EH9 3JG, Edinburgh, United Kingdom; rick.death@sac.ac.uk

In applied ethology, there are a number of occasions where a dataset contains two or more measures of the same thing from each animal and we want to determine how well they agree. Examples include testing inter- or intra-observer agreement, test-retest agreement in animal temperament, validation of automated recordings against behavioural observations, and physiological assays of duplicate samples. We can ask three questions about how well two measures agree: (1) do they differ? (bias) For example if A is usually higher than B; (2) are they associated? (how well do the A and B values correlate). In the case of categorical, ordinal or frequency data, we can also ask; (3) do they match? (how many times do A and B exactly agree). The distinction between these three concepts is often not recognised: e.g. perfect association can occur despite an absolute difference and/or no exact matching. Statistical methods that can be used to investigate agreement will be discussed: (1) for bias: e.g. Paired t or Wilcoxon tests; (2) for association: e.g. correlations; (3) for matching: e.g. % agreement, Kappa and PABAK. For data on at least an ordinal scale three or more measures can be compared using Kendall's coefficient of concordance. For data on an interval scale repeatability/reliability coefficients based on ratios of variance components will also be considered. By fitting non-linear mixed models this technique can be extended to ordinal data. These ideas will be illustrated using real data. E.g. in a study using a 0-5 locomotion score of sows with 4 observers, we identified two causes of disagreement: low scores by one observer (bias) and poor agreement at score 1 (matching). I recommend using graphical and statistical means to investigate all three kinds of agreement to enhance understanding of the nature of disagreements. I thank colleagues at Biomathematics & Statistics Scotland (BioSS) for useful discussions.

Automatic registration of grazing behaviour in dairy cows

Nielsen, Per P.[1] and Blomberg, Kristina[2], [1]Swedish University of Agriculture Sciences, Department of Animal Environment and Health, P.O. Box 7068, SE-750 07 Uppsala, Sweden, [2]Swedish University of Agriculture Sciences, Department of Animal Nutrition and Management, Kungsängens Forskningscentrum, 753 23 Uppsala, Sweden; per.peetz.nielsen@slu.se

Automated systems for monitoring behaviour of cows within dairy production have become increasingly important and the developments in technology provide new opportunities in this area. This study aimed to validate the use of a 3D activity logger (HOBO® Pendant G Data Logger), that registers the cow's head positions during grazing, to distinguish grazing behaviour on pasture from non-grazing behaviour. A total of 20 lactating dairy cows of the breed Swedish Red participated in the trial. All cows were observed each day either in the morning or afternoon. The behavioural observations were conducted during 5 hours a day for ten days, 2.5 hours in the morning (≈9:30-12:00) and 2.5 hours in the evening (≈18:00-20:30). Each cow had a logger attached to the halter and the logging interval was set to 5 seconds, which means that the head inclination was measured every 5th second. Furthermore an IceTag3D™ logger was attached to the right hind leg of each cow in order to evaluate if this information together with the information from the HOBO (IceTag recorded standing and HOBO recorded grazing then the cow was defined as being grazing) could increase the sensitivity and specificity of the experiment. The DISCRIM procedure in SAS 9.12 was used to find the optimal value of a linear discrimination between grazing and non-grazing registrations. The sensitivity of the logger was 84.6% which means that the logger recognised 84.6% of the grazing behaviour as grazing. The specificity of the logger was 79.7% which means that the logger recognised 79.7% of the non-grazing behaviour as non-grazing. When the information from the IceTag3D™ was added to the data the specificity increased to 90.2% while the sensitivity remained the same. We concluded that the HOBO logger is a useful tool for continuous automatic registration of grazing behaviour in dairy cows.

Novel object tests inform on home pen activity patterns in dairy cattle

Mackay, Jill R.D.[1,2], Haskell, Marie J.[2] and Van Reenen, Kees[3], [1]University of Edinburgh, Ashworth Laboratories, King's Buildings, West Mains Road, Edinburgh, EH9 3JT, United Kingdom, [2]SAC, West Mains Road, Edinburgh, EH9 3JG, United Kingdom, [3]Wageningen University, P.O. Box 65, 8200 AB, Lelystad, the Netherlands; jill.mackay@sac.ac.uk

At present, little information exists on how personality affects cattle behaviour outside of short term personality tests. In this study, 82 dairy cattle were fitted with IceQubes (IceRobotics Ltd., UK) to record their activity over a 62 day test period. All animals were given two combined Novel Arena/Novel Object test (NANO) lasting fifteen minutes with a two week interval between tests (average age at first test, 4.08±1.54 years; average days in milk at first test, 168.3±130.6). Variables recorded included latency to contact object, number of object contacts, disturbed behaviours shown (e.g. number of vocalisations/defecations/urinations) and locomotion duration. In GenStat (14th edition), a Principle Components Analysis based on a correlation matrix was done on the average of the variables recorded in both NANO tests, with the first component (PC1) being characterised by high locomotion and disturbed behaviours and explained 31% of the variation. PC2 was characterised by a high number of object contacts and low latency to first contact object and explained 18% of the variation. These components were termed 'anxiety' and 'neophilia' respectively. Cows were assigned a profile based on their positive or negative score in each quadrant: Anxious/Neophilic (n=17), Anxious/Neophobic (n=17), Calm/Neophilic (n=27) and Calm/Neophobic (n=21). There was no significant difference in age or days in milk at first test between the profiles. In a REML, profile had an effect on the average minimum standing bout seen in a day with both Anxious profiles having longer minimum standing bouts (F=2.97, P=0.037). Anxious/Neophobic animals tended to have a higher average maximum standing bout (F=2.31, P=0.083). However, cows which were Calm/Neophilic showed a much greater standard deviation of their average standing bout duration (F=4.23, P=0.008), and Anxious/Neophilic cows showed the smallest standard deviation. This is the first time personality in dairy cattle has been shown to have an effect on their home pen activity.

Do chopped feathers in pelleted diets of chicks reduce the appetite for feathers from conspecifics?

Kriegseis, Isabel[1], Meyer, Beatrice[2], Zentek, Jürgen[2], Würbel, Hanno[3] and Harlander-Matauschek, Alexandra[3], [1]University of Hohenheim, Department of Animal Husbandry and Breeding, Garbenstrasse 17, 70599 Stuttgart, Germany, [2]Free University of Berlin, Institute of Animal Nutrition, Brümmestrasse 34, 14195 Berlin, Germany, [3]Vetsuisse Faculty, VPHI, Division Animal Welfare, Länggassstrasse 122, 3012 Bern, Switzerland; alexandra.harlander@vetsuisse.unibe.ch

Feather-pecking laying hens eat more loose feathers than non-peckers. We hypothesized that birds provided feed containing insoluble feathers would exhibit reduced appetite for feathers and thus show reduced feather-pecking activity and a better plumage condition compared to both birds provided either with commercial food or insoluble cellulose instead of feathers. Sixty (experiment 1) and 180 (experiment 2) day-old Lohmann- Selected-Leghorn birds were divided into 12 groups of 5 (experiment 1) and 12 groups of 15 (experiment 2) birds. Four groups each had *ad libitum* access to either a pelleted commercial diet, a pelleted diet containing 5% (experiment 1) or 10% (experiment 2) of chopped feathers, or a pelleted diet containing 5% (experiment 1) or 10% (experiment 2) of cellulose over a period of 16 (experiment 1) and 17 (experiment 2) weeks. The number of severe feather-pecking bouts was recorded weekly from week 5 to 16 (experiment 1) and from week 5 to 17 (experiment 2). At the end of the experiments plumage condition per individual pullet was scored. Scores from 1 (denuded) to 4 (intact) were given for each of six body parts. The addition of 5% feathers or 5% cellulose in the diets did not reduce feather pecking or improve plumage condition. A significantly lower number of severe feather-pecking bouts in the 10% feather compared to the control groups (mean 0.8 vs. 1.02 bouts/pullet/10 min; $P<0.0129$) and an improved plumage condition of the back area in the 10% feather compared to control (mean score 4 vs. 3.3; $P<0.001$) and 10% cellulose (mean score 4 vs. 3.5; $P<0.001$) groups were recorded. The relationship between feather eating and the consequences thereof, which alter feather pecking, is unclear. Understanding this relationship might be valuable for understanding the causation of feather pecking in laying hens.

Can heterogeneity of nest-boxes reduce gregarious nesting in laying hens?

Riber, Anja B.[1] *and Clausen, Tina*[2], [1]*Aarhus University, Department of Animal Science, Blichers Allé 20, 8830 Tjele, Denmark,* [2]*Aarhus University, Department of Bioscience, Ny Munkegade 116, 8000 Aarhus, Denmark; anja.riber@agrsci.dk*

Gregarious nesting is an unwanted behaviour, where a laying hen selects an already occupied nest even though other nests are available. This may pose a financial cost to the producer and reduce animal welfare. Gregarious nesting may be caused by difficulties experienced by hens in distinguishing between nest-boxes in long rows of identical nest-boxes. It has therefore been suggested that heterogeneity of nest-boxes may reduce gregarious nesting. To test this hypothesis two experiments were performed. Six experimental groups were provided with three adjacent and different appearing nest-boxes in experiment 1, which were replaced with three adjacent nest-boxes containing different nesting-materials in experiment 2. In both experiments six control groups were provided with three adjacent and identical nest-boxes containing wood shavings. Each group consisted of 13-15 ISA Warren hens, age 27 weeks at onset of experiment 1. Daily egg collection and video recordings were performed for eight days in experiment 1 and seven days in experiment 2. The proportion of gregarious nesting was higher in the experimental groups than in the control groups (exp.1: 0.43±0.04 vs. 0.23±0.04; exp.2: 0.61±0.07 vs. 0.22±0.07; $P<0.01$). Number of visits and eggs in end-of-row nest-boxes were higher compared to those in the middle in both experimental and control groups ($P<0.05$). Number of visits and eggs were higher in standard and yellow nest-boxes compared to those with black plastic flaps in front ($P<0.05$). Furthermore, straw was preferred compared to both wood shavings ($P<0.01$) and peat ($P<0.01$) as nesting-materials. The prelaying period did not differ between the experimental or control groups ($P>0.05$). The heterogeneity in nest-box appearances and nesting-materials in these experiments did not reduce occurrence of gregarious nesting. Preferences for end-of-row nest-boxes, straw as nesting-material, and standard and yellow nest-boxes are likely to have caused the increase in occurrence of gregarious nesting in the experimental groups.

The effect of dark brooders on feather pecking on commercial farms

Gilani, Anne-Marie, Nicol, Christine and Knowles, Toby, University of Bristol, Animal behaviour and welfare group, University of Bristol, Dolberry building, Langford, Bristol, BS40 5DU, United Kingdom; Anne-Marie.Gilani@bristol.ac.uk

Commercial laying hen chicks experience periods of continuous light for up to 24 h in the first week of life, which is highly unnatural. A consequence of continuous light is that active chicks disturb sleeping ones. Crucially it also appears that feather pecking is mainly directed towards inactive birds. Previous experimental work showed reduced feather pecking in chicks brooded by dark brooders. Dark brooders provide chicks with a place to escape the light and rest undisturbed. This study aimed to extend these small-scale trials by examining the use of dark brooders on two commercial rearing farms. Each farm contributed two identical but separate houses (dark brooder versus regular brooder, brooders present for 16wks). Five rounds, each containing two flocks (10 flocks in total) of 2,000 non-beak trimmed CBT chicks were observed at 1, 8, 16 wks (rear) and two of those at 25 and 35 wks (lay) as well. Feather pecking, percentage of flock with missing feathers, individual feather scores (range 6-24) and weight were measured. Data were analysed in MLwiN, with visits as repeated measures on flocks, with treatment and age as explanatory variables. On average, across observations taken at all ages dark brooder flocks performed significantly less severe feather pecking than control flocks (0.04 versus 0.23 pecks/bird/30 min, χ^2=8.014 (df=1), P=0.005), had a significantly lower percentage of birds with missing feathers (8 versus 23%, χ^2=4.959 (df=1), P=0.03) and a significantly higher individual feather score (23.7 versus 23.5, χ^2=6.978 (1) (df=1), P=0.008). Gentle feather pecking and weight were unaffected. Weight and feather condition were significantly affected by age. This study confirmed that dark brooders can reduce feather pecking during rear, persisting into lay and may therefore prove useful in commercial pullet rearing.

A bespoke management package for loose-housed laying hens: can we mitigate against injurious pecking?

Lambton, Sarah L., Nicol, Christine J., Mckinstry, Justin L., Friel, Mary, Walton, Jon, Sherwin, Chris M. and Weeks, Claire A., University of Bristol, Clinical Veterinary Science, Langford House, Langford, North Somerset, BS40 5DU, United Kingdom; sarah.lambton@bristol.ac.uk

Injurious pecking (IP) can reduce welfare and productivity in loose-housed laying hens. This study investigated the protective effects against IP of a package of 46 on-farm management strategies (MS), determined from a systematic literature review of risk factors associated with IP. Efficacy of the package was determined by comparing levels of IP in 100 commercial flocks, each employing a variety of those MS. Rates of gentle and severe feather pecking (GFP, SFP) and vent and cannibalistic pecking (VP, CP) were recorded during 10×5 min behavioural observations of areas of 2 square metres, at 20, 30 and 40 weeks of age. Plumage scores (PD) were recorded for up to 140 birds/visit. Multilevel models created in MLwiN2.2 and Stata12.0 identified associations between IP and MSs. Flock age, beak trimming (13/100 flocks were non-beak trimmed) and strain were accounted for, where appropriate. GFP ($P=0.021$), SFP ($P=0.043$), PD ($P=0.004$) and likelihood of VP ($P=0.021$) decreased as number of MS employed increased. GFP was lower where nest boxes were not lit, and with MS designed to improve litter quality (e.g. adding highly absorbent materials or scattering grit/grain; $P<0.001$). SFP was lower where natural shelters were provided on the range ($\chi^2=4.3$, $P=0.038$) and, when groups of similar interventions were combined, SFP decreased as more MS designed to increase ranging were employed ($\chi^2=3.9$, $P=0.049$). SFP and PD were lower, and VP less likely when aerated breeze blocks were provided on slatted areas ($\chi^2=7.7$, $P=0.006$; $\chi^2=4.7$, $P=0.029$; $\chi^2=5.0$, $P=0.025$). PD was lower and VP less likely with perch heights >50cm ($\chi^2=7.2$, $P=0.028$; $\chi^2=7.4$, $P=0.025$). SFP and PD were higher where diets were mixed during changes in feed batches ($\chi^2=24.7$, $P<0.001$; $\chi^2=10.9$, $P=0.001$). Negative correlations between number of MS employed and GFP, SFP, VP and PD indicate that this package can protect against IP, with particular MS having marked impacts.

The effect of two classes of NSAIDs on the landing ability of laying hens with and without keel fractures

Nasr, Mohammed Abdel Fattah[1,2], Murrell, Jo[2], Wilkins, Lindsay[2] and Nicol, Christine[2], [1]Zagazig University, Faculty of Veterinary Medicine, Animal Wealth Development Department, Egypt, [2]Clinical Veterinary Science, Langford, University of Bristol, BS40 5DU, United Kingdom; mohammed.nasr@bristol.ac.uk

Laying hens are vulnerable to fractures of their keel bone during their laying cycle, with up to 80% prevalence of fractures in commercial systems. Our previous work revealed that hens with keel bone fractures took longer to fly down from a raised perch for a food reward than those without keel fractures, an effect that may be due to mobility restriction or pain. The administration of analgesic drugs can reveal information about the role of pain. Our initial work showed that the opioid drug butorphanol decreased the latency to land in hens with keel fractures. The aim of the current study was to examine whether non-steroidal anti-inflammatory drugs (NSAIDs) (carprofen and meloxicam) would have a similar effect. We collected 103 Lohman Brown hens (35-38 weeks of age) from commercial farms and found by palpation that 51 had suspected old keel fractures and 52 had no suspected fractures. The latency to fly down from different perch heights (50, 100 and 150 cm) to the floor for a food reward was recorded for each hen given each drug (25 mg/kg carprofen or 5 mg/kg meloxicam s/c), and for each hen under a saline control through a randomized cross over design. Unexpectedly, carprofen increased the latency to land from different perch heights in hens with keel fractures. But meloxicam decreased the time taken by hens with keel fractures to land from the highest perch compared with saline injection (2.67±0.58 vs. 3.20±0.47 P=0.30 for 50 cm; 47.84±13.14 vs. 57.20±14.65 P=0.16 for 100 cm; 92.05±21.22 vs. 108.92±20.23 P=0.04 for 150 cm). It appears that meloxicam has an analgesic effect on hens with keel bone fractures, suggesting that hens with old keel fractures are in some degree of pain.

Reducing stocking density improves mating behaviour in broiler breeders

De Jong, Ingrid, Gunnink, Henk, Van Emous, Rick and Lourens, Sander, Wageningen UR Livestock Research, P.O. Box 65, 8200 AB Lelystad, the Netherlands; ingrid.dejong@wur.nl

Broiler breeder males often show rough behaviour towards females during mating, which may cause fear and feather and skin damage in the females. Our hypothesis was that the high stocking density applied under commercial conditions hampers correct learning and performance of mating behaviour. Broiler breeders (Ross 308) were housed at standard stocking density (14 females/m^2 or 8 males/m^2 during rearing, 8 birds/m^2 during production) or low stocking density (7 females/m^2 or 5 males/m^2 during rearing, 5 birds/m^2 during production) from day 1 to 60 weeks of age. Standard management was applied. During production there were 10% males per group and group size was 150 birds. The experiment was of a 2×2 factorial design, resulting in 4 treatment groups: SL (standard stocking density during rearing, low stocking density during production), SS, LS and LL; n=8 groups/treatment. Behaviour and plumage condition were scored during rearing and production. In particular stocking density during production affected mating behaviour. LL and SL groups had more successful copulations at all ages (e.g. 0.51, 0.33, 0.22, 0.21 successful copulations/male/5 min at 30 weeks for LL, SL, SS, LS; $P<0.05$), more copulations were preceded by courtship behaviour at all ages (e.g. 32%, 23%, 8%, 15% at 30 weeks for LL, SL, SS, LS; $P<0.001$), and hens crouched significantly more in response to the male at all ages (e.g. 27%, 20%, 4%, 4% copulations with crouching at 30 weeks for LL, SL, SS, LS; $P<0.001$). During production LL and SL hens showed less struggling during mating as compared to the SS and LS groups at all ages ($P<0.05$), and LL and SL groups had a better plumage condition at all ages than SS and LS groups ($P<0.001$). Thus, reducing stocking density during production significantly improves mating behaviour resulting in a better plumage condition in the hens.

Individuality of ranging behavior in large flocks of laying hens

Gebhardt-Henrich, Sabine G. and Froehlich, Ernst K.F., Center for proper housing: poultry and rabbits, Burgerweg 22, 3052 Zollikofen, Switzerland; sabine.gebhardt@bvet.admin.ch

Laying hens with outdoor access are often kept in flocks of several thousand birds. In this study we investigated individual variation in different traits of ranging. Five percent of hens (9-15 months of age) randomly selected from twelve commercial flocks of 2,000 to 18,000 birds were fitted with RFID tags and their movements between house, porch, and pasture were recorded for at least eleven days. At least 48% of the hens were registered on pasture. The analyses focus on these birds. Total duration spent on pasture/day, number of days when pasture was entered, and the number of days when the porch was entered from the house had significant bimodal distributions (coefficient of bimodality >0.555, duration: sign test: M=5, P=0.006, number of days on pasture: sign test: M=6, P=0.0005, number of days on the porch: sign test: M=5, P=0.002, n=10). Daily duration on pasture was consistent among individuals (mean r_s=0.54±0.04, P<0.0001, n=12 flocks) and correlation coefficients did not decrease over days. Daily duration and number of days using the pasture were positively correlated: r_s=0.62±0.06, P<0.0001, n=12 flocks. These parameters were influenced by flock size: Hens in flocks >9,000 spent more days on pasture (χ^2_2=7.85, P=0.02) for a longer time (χ^2_2=8.15, P=0.02). The earlier in the day the hens went out on pasture, the more time they spent there (r_s=- 0.55±0.03, P<0.0001, n=12 flocks). Thus, there seem to be different categories of hens regarding use of pasture. Although some hens hardly ever used the pasture, there were hens in each flock using it regularly.

Behavioural organisation and the reward cycle

Seehuus, Birgitte and Blokhuis, Harry, Swedish University of Agricultural Sciences, Animal Environment and Health, Box 7068, 75007, Sweden; birgitte.seehuus@slu.se

The reward cycle is a theoretical model that aims to make connections between behavioural stages and related affective states: an appetitive motivational stage linked to the seeking/ wanting of a resource, a consummatory motivational stage linked to sensory pleasure and a post-consummatory motivational stage associated with satiety/relaxation. Since these affective states are considered positive, they have attracted recent attention due to the relevance of positive emotions in the animal welfare discussion. Although these three motivational stages are often taken for granted, there is to our knowledge no data to support a greater coherence of behaviours within stages than between stages. We studied feeding behavior in chicks to see if observed behavioural sequences supported a separation of the three stages as a first step to further develop the feed reward cycle model. Two batches of 18 chickens were kept in three pens from the day of hatching until the end of the experiment. In week six the pens were video recorded and one hour of recording from each pen and batch were chosen for continuous observations. The behavior of a focal animal was recorded every 10 seconds and probabilities of behavioural sequences were calculated. We identified coherent behavioural sequences related to appetitive (pecking/scratching), consummatory (feeding) and post-consummatory (preening/resting) stages. The behaviours corresponding to the appetitive stage (pecking – pecking and pecking /scratching – scratching/pecking) had a transition probability of 0.083 and 0.01 respectively; the transition probability for keeping on feeding was 0.166. The probability for keeping on preening was 0.158, for resting 0.044 and for switching between preening and resting 0.016. The probabilities for moving between stages were lower. Data suggest a higher number of transitions between behaviours related to an appetitive, consummatory and post-consummatory stage respectively. This supports the separation of these three stages in the organization of feeding behavior.

Investigating the reward cycle for play in lamb

Lidfors, Lena and Chapagain, Durga, Swedish University of Agricultural Sciences, Department of Animal Environment and Health, P.O. Box 234, 532 23 Skara, Sweden; Lena.Lidfors@slu.se

The expression of play in farm animals could be an indicator of positive emotions, but could this be studied by using a reward cycle for play? The aim was to test if lambs would show behaviours indicative of anticipation, consumption and relaxation in an arena with play objects.10 pairs of male lamb of Swedish Fine Wool and Dorset were housed in litter pens (2×3 m) at Götala Research Farm in Sweden. Pairs of lambs were taken to a play arena three times a week during five weeks (first week for learning). They were kept in a holding pen where they could look into the play arena for 5 minutes and thereafter released for 20 minutes in the play arena (5.9×5.5 m) which had a ball, two chains and a tunnel. Thereafter they walked back to their home pen and were observed for 6 minutes. Observations were made as frequency per minute in all three places. Wilcoxon signed ranks test was used for pairwise comparisons. Lambs showed more anticipatory behaviours (3.95±0.17) than non-anticipatory behaviours (1.34±0.05) in the holding pen ($P<0.01$). In the play arena play behaviours (4.05±0.29) was higher than other behaviours (2.62±0.10, $P<0.01$). Social play (1.88±0.21) was performed more than locomotor (1.06±0.15) and object play (1.10±0.09). The level of behaviours indicating relaxation (1.08±0.039) was not higher than behaviours indicating no relaxation (0.95±0.10) in the home pen afterwards (n.s.). During the test anticipatory behaviours did not change, whereas play behaviour decreased and relaxation behaviours increased. It is concluded that lambs seem to show behaviours indicating anticipation to play and they do play in a known play arena, but observation time was too short to detect relaxation after play.

Investigating behavioural indicators of positive and negative emotions and emotional contagion in pigs

Reimert, Inonge[1], Bolhuis, J. Elizabeth[1], Kemp, Bas[1] and Rodenburg, T. Bas[2], [1]Wageningen University, Adaptation Physiology Group, De Elst 1, 6708 WD Wageningen, the Netherlands, [2]Wageningen University, Animal Breeding & Genomics Centre, De Elst 1, 6708 WD Wageningen, the Netherlands; inonge.reimert@wur.nl

Studies providing indicators of positive emotions in farm animals are, still, scarce. We investigated therefore behavioural indicators of positive and negative emotions in pigs. Two test pigs of a pen (n=6 pens) were, on 7 days divided over a 10-day period, brought to a test room twice per day where they were exposed for 5 min to a positive (pairwise access to a large compartment with substrates) and a negative treatment (social isolation in a small compartment combined with negative, unpredictable interventions) in a random order. Behaviours were scored from video using focal sampling and continuous recording. Data were analysed using mixed models or the Fisher's Exact Test. During the positive treatment, pigs showed more play (89% occurrence vs. 0%, $P<0.001$), barks (28% occurrence vs. 0%, $P<0.05$) and tail wagging ($6.3\pm1.3\%$ of time vs. $0.2\pm0.1\%$, $P<0.001$) compared to the negative treatment, where pigs showed e.g. more defecating (1.0 ± 0.1 occurrences/min vs. 0.3 ± 0.1, $P<0.001$) and high-pitched vocalisations (100% occurrence vs. 0%, $P<0.001$). After the 10-day period, two pen mates which were naive to the treatments were brought along with the test pigs to the test room and placed in an adjacent compartment were no treatment was given. Both test and naive pigs played more during the positive treatment of the test pigs (test pigs: 100% vs. 0%, $P<0.01$, naive pigs: 83% vs. 0%, $P<0.05$) and showed more defecating during the negative treatment of the test pigs (test pigs: 1.2 ± 0.1 vs. 0.4 ± 0.1, $P<0.05$, naive pigs: 0.9 ± 0.3 vs. 0.6 ± 0.3, $P<0.1$). In conclusion, we could confirm play and barks as indicators for positive emotions, and we suggest tail wagging as a new indicator. Furthermore this study suggests that emotional contagion exists in pigs, because play and defecating seem to stimulate play and defecating in others and thereby also positive and negative emotions in others.

Does training for anticipation of a positive reinforcement lead to a long term positive emotional state in growing pigs?

Imfeld-Mueller, Sabrina[1], Otten, Winfried[2], Bruckmaier, Rupert M.[3], Gygax, Lorenz[4] and Hillmann, Edna[1], [1]Animal Behaviour, Health and Welfare Group, Institute of Agricultural Sciences, ETH Zurich, Universitaetsstr. 2, 8092 Zurich, Switzerland, [2]Leibniz Institute for Farm Animal Biology, Research Unit Behavioural Physiology, FBN Dummerstorf, Wilhelm-Stahl-Allee 2, 18196 Dummerstorf, Germany, [3]Veterinary Physiology, Vetsuisse Faculty, University of Bern, Bremgartenstrasse 109a, 3001 Bern, Switzerland, [4]Centre for Proper Housing of Ruminants and Pigs, Federal Veterinary Office, Agroscope Reckenholz-Tän, Tänikon, 8356 Ettenhausen, Switzerland; sabrina-imfeld@ethz.ch

Our aim was to determine whether a limited but regularly provided food ball filled with maize leads to changes in behavioural and physiological reactions to stressful situations in pigs. Seventy-two growing pigs were assigned to groups of three and to three different treatments. In the 'anticipated enrichment' (aE) and the 'enrichment only' (Eo) treatments, the ball was provided about 10 times per week for 10 minutes each during the fattening period. In the aE-treatment, the ball was given 30 s after a tone sequence was played. In the Eo-treatment, the ball was provided without temporal link to the acoustic signal. Pigs in the 'control'-treatment (C) never received the ball. At the age of 14-15 weeks, a combined open-field and novel-object test was conducted with one animal per group. Behaviour during the test was analyzed, and saliva samples were taken to measure basal cortisol levels. At slaughter, plasma catecholamines, and cortisol were measured. In samples from the ventral tegmental area (VTA), nucleus accumbens, and prefrontal cortex of two pigs per group, neurotransmitter concentrations were measured. Data were analysed with linear mixed-effects models. During open-field, the only treatment effect was the higher probability of a pig screaming or squeaking for Eo-pigs (P=0.001; 1 compared to aE- (half the pigs) and C-treatment (1/4 of the pigs)). The DOPAC/dopamine ratio as an indicator for presynaptic neuronal activity in VTA (P=0.10), considered to indicate better coping capacities, basal cortisol (P=0.10), and plasma cortisol (P=0.09) tended to be highest in the aE, followed by the Eo and C treatments, suggesting a decreased HPA-activity and a sign for reduced welfare in C-pigs. Even if treatment differences were small, probably due to quite enriched housing conditions of all treatments, results indicate the potential to induce increased welfare in growing pigs by using positive anticipation.

What are positive feelings: examples from achievement excitement, positive cognitive bias and guilt

Broom, Donald, University of Cambridge, Veterinary Medicine, Centre for Animal Welfare and Anthrozoology, Department of Veterinary Medicine, University of Cambridge, Cambridge CB3 0ES, United Kingdom; dmb16@cam.ac.uk

The word positive, in this context has several facets to it, depending on the background of the researcher. For some, positive means: increasing fitness, eliciting approach, associated with an action likely to be replicated, acting as a positive reinforcer, or associated with physiological changes such as oxytocin increase, vagal nerve activity or some kinds of opioid activity. A feeling is a brain construct, involving at least perceptual awareness, which is associated with a life regulating system, is recognisable by the individual when it recurs and may change behaviour or act as a reinforcer in learning. An emotion is the physiologically describable aspects of this. When animals attain a physical objective or learn something, there may be measurable excitement. Examples of this described here include our studies of the behavioural and physiological changes that occur at the point of learning in young cattle and sheep. These changes in the animal would seem to be evidence of emotions/feelings and to be positive according to most criteria. If an individual is acting in a way that indicates that it is going for the positive in an ambivalent situation, is this evidence of a positive emotion? Although it may well be, some of the problems with interpretation of cognitive bias will be discussed. The work of Carla Torres Pereira describing the behavioural and heart-rate changes in a situation in which a dog has done something that it was instructed not to do is presented and the concept of guilt described. Is guilt all negative? Is pain all negative? Can sexual pleasure/achievement pleasure/eating pleasure have negative consequences? Information of any kind is often valuable, whether the feeling is positive or negative. Questions about function and value may be closely or more distantly be related to immediate feeling.

Goat kid calls differ according to emotional intensity

Briefer, Elodie F. and McElligott, Alan G., Queen Mary University of London, Biological and Experimental Psychology Group, School of Biological and Chemical Sciences, Mile End Road, London E1 4NS, United Kingdom; e.briefer@qmul.ac.uk

Public concern about animal welfare is strongly based on the attribution of emotional states to animals. However, techniques to assess emotions are lacking. This study was aimed at finding convenient and non-invasive vocalization-linked methods to assess emotions in goats. The emotional state of a caller causes changes in its muscular tension and action of the vocal apparatus, which impact on the structure of vocalisations. We conducted a two-part experiment with 13 goat kids (10-18 days old), in which they were separated for a short time from their mothers. We recorded and analysed the calls they produced. In the first treatment, kids could interact vocally and visually, but not physically with their mothers. In the second treatment, kids were visually isolated from their mothers, and calls of their mothers were played back from a loudspeaker. Therefore, in this situation, kids could hear, but not see their mothers. These two treatments were expected to trigger negative emotions of different intensities (i.e. higher intensity in the second treatment). We measured 32 acoustic parameters in 8 calls per kid per treatment. We used GLMM, controlled for age and sex, and with the identity of kids nested within the identity of their mothers as a random structure, to compare the vocal parameters. Our results revealed changes in several vocal parameters between the first and second treatments, including an increase in duration (t_{190}=2.29, P=0.023), fundamental frequency (GLMM: t_{190}=4.15, P<0.001) and frequency modulation (t_{182}=3.44, P<0.001), and a decrease in amplitude modulation (t_{190}=-3.19, P=0.002) and formant values (e.g. first formant: t_{190}=-3.61, P=0.0004). Furthermore, we suggest techniques that could be used to measure behavioural, vocal and physiological responses to situations triggering different emotional valence (positive or negative) and intensity, in goats and other mammals. These non-invasive indicators of emotions could be very useful for assessing welfare.

Acoustic properties of piglet voices determine what emotional intensity and valence people attribute to them

Špinka, Marek[1], Maruščáková, Iva[1,2], Linhart, Pavel[1,3] and Tallet, Céline[1,4], [1]Institute of Animal Science, Department of Ethology, Přátelství 815, 104 00 Prague - Uhříněves, Czech Republic, [2]Charles University, Department of Zoology, Viničná 7, 128 00 Prague, Czech Republic, [3]University of South Bohemia, Department of Zoology, Branišovská 31, 370 05 České Budějovice, Czech Republic, [4]INRA, UMR PEGASE, PEGASE, 35590 Saint-Gilles, France; spinka.marek@vuzv.cz

Vocally expressed emotions can be sometimes recognised by another species. This understanding may rely on specific acoustic parameters of the calls. We investigated how humans perceive emotions in piglet voices and how they assess the type of situation from the calls dependent on the vocalisations' acoustic parameters. 48 examples of calls, each by a different 7-14 days old piglet (12 examples recorded in each of 4 situations, namely castration, isolation, reunion with mother and nursing) were played back to 61 university students who had very limited previous experience with live pigs or pig vocalisations. Each respondent judged 12 examples (3 per situation, without knowing about the situations) for emotional intensity, 12 examples for valence and decided for another 12 examples in which situation they had been recorded. The examples were analysed for their acoustic properties, including call rate, proportion time spent calling, pitch, harmonic-to-noise-ratio and jitter. In a mixed linear model that included random effects of person identity and call identity, the acoustic properties explained >50% of judged intensity and valence. Estimated emotion intensity increased with the pitch of the call and with the proportion of time spent vocalising. Valence was judged more positive with decreasing pitch, decreasing time spent vocalising and increasing vocalisation rate ($P<0.001$). People identified situations correctly in 51%, significantly ($P<0.0001$) above chance level of 25%. Interestingly, correct assignment was highly correlated with judged intensity. For castration and reunion calls, the higher intensity the respondents assigned to an example, the more likely the situation was correctly identified ($r=0.77$, $P<0.01$ and 0.69, $P<0.05$, respectively). For nursing calls, calls perceived as less intense were more often correctly assigned ($r=-0.821$, $P<0.01$). In conclusion, people judge emotions from pig calls using simple acoustic properties and perceived intensity may guide judgements about the situation.

Morphine induces an optimistic judgement bias after consumption of a palatable food reward in sheep

Verbeek, Else[1], Ferguson, Drewe[1], Quinquet De Monjour, Patrick[2] and Lee, Caroline[1], [1]CSIRO, Livestock Welfare, Locked Bag 1, Armidale NSW 2350, Australia, [2]ParisAgroTech, 16 rue Claude Bernard, 75005 Paris, France; else.verbeek@csiro.au

Positive emotions are relevant to animal welfare, yet their regulation is poorly understood. We hypothesised that an opioid receptor agonist (morphine) vs. an opioid receptor antagonist (naloxone) would alter judgement bias in sheep and that their response would be influenced by whether they received a palatable (food) or unpalatable (wood chips) reward prior to the judgement bias test. Sheep were trained in an arena to approach a positive location associated with conspecifics and not approach a negative location associated with a dog. Three non-trained, non-reinforced ambiguous locations were situated between the positive and negative locations. Following training, sheep were randomly allocated to one of three treatments (n=10 per treatment): morphine (1 mg/kg), naloxone (2 mg/kg) and control (sterile water). Within each treatment, half the ewes received a small food reward on day (d) 1 and wood chips on d2 and the other half wood chips on d1 and food on d2 before exposure to each location. Judgement bias was assessed 10 min after drug administration by recording the latency to approach each of the five locations. Data were analysed by REML and post-hoc orthogonal contrasts. There was a treatment × reward type × day interaction ($P<0.05$): this effect was mostly due to the reduced latency to approach the locations in the morphine ewes that received the food reward on d1 (morphine: 10.2±3.2s vs. control: 16.3±2.7s and naloxone: 14.1±2.6s, $P<0.05$). No treatment differences were found after presentation of the wood chips on both days. In conclusion, the results of d1 suggest that morphine induced an optimistic judgement bias after consumption of a food reward, but morphine did not alter judgement bias after receiving the wood chips. Naloxone had no effect on judgement bias. Our results contribute to the evidence that the opioid system is involved in modulating positive emotions.

A glass full of optimism: enrichment effects on cognitive bias in a rat model of depression

Richter, Sophie H.[1], Hoyer, Carolin[1], Schick, Anita[2], Gass, Peter[1] and Vollmayr, Barbara[1], [1]Central Institute of Mental Health, Medical Faculty Mannheim, Heidelberg University, Animal Models in Psychiatry, J5, 68159 Mannheim, Germany, [2]Central Institute of Mental Health, Medical Faculty Mannheim, Heidelberg University, Department of Cognitive and Clinical Neuroscience, J5, 68159 Mannheim, Germany; helene.richter@zi-mannheim.de

Traditionally, there has been a strong emphasis on the study of negative affective states in animal welfare science. However, it is now widely accepted that both the absence of negative states and the presence of positive emotions contribute to good welfare, making it indispensable to analogously explore positive states. Using a spatial judgement task we aimed to investigate affective state in a genetic rat model of depression. Cognitive bias was assessed prior to and following environmental enrichment to assess whether this induces an optimistic shift in emotional state, thereby ameliorating depression-like symptoms in our rat model. Twenty-four males of congenitally helpless (cLH, n=12) and non-helpless (cNLH, n=12) rats were tested twice in a spatial judgement task, half of which being transferred to enriched housing conditions one week before the second test commenced. To control for differences in reward sensitivity between cLH and cNLH rats, we additionally tested for hedonic-like behaviour by measuring the consummatory response to sweetened condensed milk (SCM). Transferring animals into enriched cages resulted in reduced latencies to choose ambiguous locations (2-way repeated measures ANOVA, $F_{1,18}$=8.494, P=0.009), implying a more optimistic interpretation of ambiguous spatial cues in enriched housed animals. While this effect was present in both strains, SCM intake was only increased in cNLH rats (T test, cNLH: T=-2.328, P=0.042; cLH: T=1.355, P=0.205), indicating that cognitive appraisal and reward sensitivity in our animals are differentially affected by enrichment. Taken together, our results demonstrate that environmental enrichment can promote an optimistic judgement bias, possibly indicative of a more positive affective state. This underscores the importance of species-appropriate housing conditions that provide adequate sensory and motor stimulation.

An active-choice judgement bias task for pigs: a comparison between Göttingen minipigs and conventional farm pigs

Murphy, Eimear, Nordquist, Rebecca and Van Der Staay, Franz J., Utrecht University, Farm Animal Health, Yalelaan 7, 3584 CL Utrecht, the Netherlands; E.M.Murphy@uu.nl

Mood state can influence the judgement of ambiguous stimuli as positive or negative. This research aimed to develop an active-choice judgement bias task for pigs, and to compare performance across breeds. Eight Göttingen minipigs and seven conventional farm pigs were trained on a conditional discrimination task, where tone-cues signaled the location of large and small rewards (left/right goal-pot). Incorrect choices were unrewarded and resulted in a time penalty. After learning, pigs were tested over four sessions where three ambiguous tone-cues were presented amongst the learned-cues per session. Ambiguous cues were neither rewarded nor given a time penalty. Choice of goal-pot was recorded. After a break, pigs were retrained and tested again. Minipigs learned the initial discrimination faster than conventional pigs (8.38±0.80 vs. 13.14±0.67 sessions; t_{13}=4.49, P=0.0006), but there was no difference between breeds in relearning. In test sessions both breeds responded faster to the positive-cue than the negative-cue (6.78±0.23s vs. 9.18±0.56s; $F_{1,13}$=55.38, P<0.0001) confirming a preference for the large reward. An 'Optimistic Score' (OS) was calculated as the percentage of choices for the positive goal-pot in ambiguous trials, per cue-type and in total. In both tests, there were no breed differences in total OS (mini: 48.81±8.42%, conv. 29.17±9.18%; $F_{1,13}$=2.09, P=0.17) nor per cue-type. In retesting both breeds chose negative more often in ambiguous trials ($F_{1,13}$=18.53, P=0.0009). Pigs of different breeds easily learned this task, and responded as expected to the ambiguous cues: the closer it was to the positive cue, the higher was the OS in both breeds of pigs (Test 1: $F_{2,28}$=11.67, P=0.0002; Test 2: $F_{2,28}$=6.79, P=0.0105). However, over-presentation of the ambiguous cues may lead to some extinction of responding to ambiguous cues. The potential for studying both positive and negative emotions through a judgement bias task is discussed.

Design of cognitive bias studies – review of current research and a pilot study in domestic pigs

Düpjan, Sandra[1], Ramp, Constanze[2], Tuchscherer, Armin[3] and Puppe, Birger[1,2], [1]Leibniz-Institute for Farm Animal Biology, Research Unit Behavioural Physiology, Wilhelm-Stahl-Allee 2, 18196 Dummerstorf, Germany, [2]Rostock University, Faculty of Agricultural and Environmental Sciences, Justus-von-Liebig-Weg 6, 18059 Rostock, Germany, [3]Leibniz-Institute for Farm Animal Biology, Research Unit Genetics and Biometry, Wilhelm-Stahl-Allee 2, 18196 Dummerstorf, Germany; duepjan@fbn-dummerstorf.de

Cognitive bias, especially judgement bias (i. e. emotionally biased evaluation of ambiguous stimuli) might be a proxy measure for affective valence in farm animals. We critically review studies on cognitive bias in animals, emphasizing variables of study design, namely type of reinforcer, stimulus, and task. Subsequently, we present a pilot study on judgement bias in domestic pigs, using a spatial go/no-go task with accessible food as positive reinforcer and inaccessible food as negative reinforcer. 32 German Landrace pigs (6-7 weeks old) were trained to discriminate a rewarded (R) from an unrewarded (U) trough position (5 days: 6×R, 6×U). After training, half of the animals were subjected to repeated social isolation (ISO; 3 days: 2×2 h; testing period: 1×2 h), while the other half stayed in their pens (CTRL). In the testing period (3 days), the trough was presented at R, U and 3 equidistant, intermediate positions (near rewarded NR, middle M, near unrewarded NU). Latency to open the trough and general behaviour were recorded during 2 minutes. Salivary cortisol was measured at 8 a.m. (basal levels), and directly before and after testing. Data were analysed using mixed effect models (including treatment, position, day and subject) and pairwise t-tests. Latencies towards NR (LSM±S.E.=1.25±5.48 s), M (5.24±5.33 s), and NU (7.75±5.33 s) did not differ from R (1.98±2.36 s), while all of these differed from U (43.11±2.64 s), indicating general optimism. ISO and CTRL did not show different basal salivary cortisol levels (e.g. testing day 3: ISO=1.92±0.31 ng/ml, CTRL=2.02±0.31 ng/ml), cortisol responses to testing, latencies to test positions, and other behaviours. We conclude that repeated social isolation did not have measurable effects on the subjects' affective state as indicated by behaviour, HPA-reactivity, and evaluation of ambiguous stimuli. Hence, at least in pigs further validation of the cognitive bias approach is required.

Validating a new behavioral measure of welfare and its relationship to theoretical models of welfare

Franks, Becca, Higgins, E. Tory and Champagne, Frances A., Columbia University, Psychology, 1190 Amsterdam Ave MC 5501, New York City, NY 10027, USA; beccafranks@gmail.com

Measures of pessimistic cognitive bias have been found to indicate poor welfare, but current measures do not capture the intersection between cognitive bias, motivation, and other behaviors associated with welfare. To address this limitation, we developed and validated a modified cognitive bias test. Group-housed rats (n=60) were trained in four arms of an eight-arm-radial-maze. They learned to expect successes in two arms (a treat-arm and a darkness-arm) and failures in the other two (a nontreat-arm and loud-noise-arm); arms were presented simultaneously. After training, 40 animals were isolated for two weeks. Then, all 60 animals were re-tested in the maze. For the first four minutes, they had access to the previously exposed arms, but in the final minute, they also had access to the previously unexposed arms. In the final minute, isolated animals took longer to begin exploring the new arms (P=0.04), explored less thoroughly (P=0.06), spent less time exploring (P=0.01), and spent more time in the familiar-safe parts of the maze (P=0.004). Cronbach's alpha for the four exploration indicators was 0.75, suggesting they capture a single latent variable. Using a composite score as an indicator of overall welfare we confirmed that welfare decreased in the isolated animals (P=0.03). As predicted, regardless of isolation condition, this welfare measure was not associated with darkness achieved in the first four minutes, but was associated with fewer treat activations in the first four minutes (P=0.02) and with more uneaten treats by the end of five minutes (P=0.002). Capturing several behaviors and motivations related to welfare beyond a standard cognitive bias test, this new measure of welfare demonstrates reliability and has predictive, construct, and discriminant validity. As such, it may help explicate several theoretical models of welfare.

Housing conditions and cognitive bias in Japanese quail

Horváthová, Mária[1], Pichová, Katarína[1,2] and Košťál, Ľubor[1], [1]Institute of Animal Biochemistry and Genetics, Slovak Academy of Sciences, Department of Physiology and Ethology, Moyzesova 61, 90028 Ivanka pri Dunaji, Slovakia (Slovak Republic), [2]Faculty of Science, Comenius University, Department of Animal Physiology and Ethology, Mlynská dolina, 84215 Bratislava, Slovakia (Slovak Republic); Lubor.Kostal@savba.sk

Cognitive bias represents an innovative indicator of animal emotion and welfare. Judgement bias reflects the tendency of a subject to show behaviour indicating anticipation of either positive or negative outcomes in response to affectively ambiguous stimuli. We have adopted this concept for the assessment of Japanese quail welfare in relation to housing conditions. Cognitive performance of quail was measured using an operant chamber with touch-screen monitor. In two experiments (n=6 and n=12) adult hens were first trained in a Go/NoGo task to respond by pecking to a positive stimulus (white circle) associated with a positive event (access to feeder) and to refrain from pecking at the negative stimulus (80% grey circle) to avoid a negative event (70 dB white noise). Hens were kept in cages and food restricted in order to increase their feeding motivation. Once the birds discriminated successfully both stimuli (significant r in 3 consecutive daily sessions), they were within each experiment divided into the two groups according to housing conditions (wire cage vs. deep litter) and after adaptation they were subjected to ambiguous stimuli (colours between positive and negative stimuli) test in 3 independent sessions. The proportion of birds classifying ambiguous stimuli as positive was higher in deep litter housed birds than in caged ones, as confirmed by the three way ANOVA (the effect of housing conditions – $P<0.01$ in Exp1 and $P<0.001$ in Exp2; effect of stimulus – $P<0.001$ in both Exp1 and Exp2; effect of testing day – $P<0.001$ in Exp1; effect of interaction of housing conditions and testing day – $P<0.05$ in both Exp1 and Exp2). To conclude, these pilot studies suggest that the cognitive bias concept can represent a suitable method for the assessment of affective states in poultry. This work was supported by APVV-0047-10 and VEGA-2/0192/11.

Negative stimuli affect the performance of sheep in a spatial cognitive task

Doyle, Rebecca[1], Freire, Rafael[1] and Lee, Caroline[2], [1]Charles Sturt University, School of Animal and Veterinary Science, Boorooma St, Wagga Wagga 2678, Australia, [2]CSIRO, Livestock Industries, Locked Bag 1, Armidale 2350, Australia; rdoyle@csu.edu.au

Moods and emotion-eliciting stimuli can impact cognitive performance. Being able to measure how experiences and stimuli alter cognition can therefore give insight into an animal's affective state. This study aimed to investigate how spatial cognition of sheep was affected when exposed to a negative, novel stimulus (white noise). It also aimed to investigate how this was influenced by positive and negative affective states. Sixty-four castrated male lambs were habituated to a yard leading to a maze designed to test spatial cognition, with the aim to induce differing affective states prior to maze testing. Positive Habituation (and assumed positive affective state, n=33) used feed and conspecifics as reinforcers of the environment, Negative Habituation involved isolation and the presence of a dog (n=31. Following this, sheep traversed a maze individually whilst being exposed to white noise (positive habituated n=17, negative n=15) or no noise (positive n=16, negative n=16). Total time to complete the maze and duration of errors were measured for three consecutive days. Results were then analysed using LMMs and included interactions of Day, Habituation and Noise in the maze as factors, and Sheep as a random factor. White noise tended to increase total time taken to complete the maze and the duration of errors made on Day 1 (145s vs. 110s, P=0.075 and 51s vs. 40s, P=0.053 respectively). Negative habituation also increased the duration of errors made on Day 1 (77s vs. 63s, P=0.045). Both negative stimuli and the habituation treatment affected the innate problem solving skills of sheep in spatial tasks, but did not affect performance thereafter. These results suggest that cognitive task performance may be a useful measure of affective state and stimulus valence in sheep.

Factors influencing injuries and social stress in horned and dehorned goat herds

Waiblinger, Susanne, Nordmann, Eva, Mersmann, Dorit and Schmied-Wagner, Claudia, University of Veterinary Medicine, Institute of Animal Husbandry and Welfare, Veterinärplatz 1, 1210 Wien, Austria; Susanne.Waiblinger@vetmeduni.ac.at

Disbudding of kids is common in intensive dairy goat farming argued by a high risk of injuries in horned herds. We aimed at evaluating social stress and injuries in dairy goat herds with or without horned animals and at identifying risk factors. On 45 farms (78-518 lactating goats) social behaviour of goats was observed directly for 3×2 h distributed over two days (barn divided into segments for continuous behaviour recording), and goats were examined for injuries. Housing, management and human-animal-relationship (e.g. behaviour of milker, goats' reactions to humans) were recorded in detail. 15 farms had purely hornless herds, i.e. genetically hornless and disbudded goats (=dehorned), 30 herds were mixed of horned and hornless goats (=horned). Comparing horned and dehorned herds by Mann-Whitney-U test revealed no difference (P>0.05) in the number of agonistic interactions with body contact (AgoBC) ranging from 0.13-1.33 interactions/goat*10 min. Prevalence of injuries at the udder was higher in horned herds (MWU: P<0.01), but variation was huge (horned: median, min-max: 0.43, 0.05-1.71 injuries/goat, dehorned: 0.22, 0.02-2.01). There were no further differences. Linear regression was used to identify influencing factors. Proportion of horned animals (range 0-78%) was no predictor in regression models for any of the variables. AgoBC was lower (model: R^2adj.=0.61, P<0.001, n=45) when kids stayed longer with their mothers, with fewer goats per drinker, better quality of roughage, longer working time close to goats (all P<0.01), well-structured housing (P>0.1). Injuries at the udder (horned herds) were lower with more space at the feeding place, lower competition at drinkers and roughage, no purchase of female goats, fewer number of milkers, farmers adopting special management for horned goats (all P<0.01, R^2adj.=0.66, P<0.001, n=30). Farms with low prevalence of injuries and aggression existed regardless of horns. Improving housing and management enables keeping of horned dairy goats without higher social stress or occurrence of injuries. We acknowledge funding by BMLFUW and BMG, Project-Nr.100191.

Disbudding of calves: the effects of an oral sedative agent on behavioural sedation scores

Hokkanen, Ann-Helena[1,2], Raekallio, Marja[2,3], Salla, Kati[2,3], Hänninen, Laura[1,2], Viitasaari, Elina[1,2], Norring, Marianna[1,2], Raussi, Satu[2] and Vainio, Outi[2,3], [1]University of Helsinki, Faculty of Veterinary Medicine, Department of Production Animal Medicine, P.O. Box 57, 00014 University of Helsinki, Finland, [2]University of Helsinki, Research Centre for Animal Welfare, P.O. Box 57, 00014 University of Helsinki, Finland, [3]University of Helsinki, Faculty of Veterinary Medicine, Department of Equine and Small Animal Medicine, P.O. Box 57, 00014 University of Helsinki, Finland; ann-helena.hokkanen@helsinki.fi

Administration of local anaesthetic (LA) prior disbudding improves calf welfare and usually requires sedation. However, in many European countries pain relief is not mandatory for calves disbudded before four weeks of age, and only veterinarians are allowed to use injectable sedatives. There is an oro-transmucosal sedative (Domosedan Gel® 7.6 mg/ml oromucosal gel, Orion Pharma) available for horses to be used by owners themselves, but it has not been studied on calves. We used 20 routinely disbudded calves for the study. We sedated 10 calves with Domosedan Gel® (detomidine, 80 µg/kg) (GEL) and 10 calves with intravenous Domosedan® (detomidine, 30 µg/kg) (IV). All calves got anti-inflammatory drug (meloxicam) subcutaneously. Prior disbudding (heat cauterization) LA (lidocain) was administered subcutaneously on top of corneal nerve and around horn buds. We measured heart rate (HR), and behavioural sedation scores (CSS: from 0 fully awake to 16 deeply sedated) including palpebral reflex, eye position, jaw and tongue relaxation, resistance to positioning in lateral recumbency and general appearance. When calves were visually sedated, we estimated the correct timing for LA with a needle test. We analyzed the effect of sedatives on parameters with linear mixed models, taking repeated measurements into account. The study was approved by Animal Ethics Committee. Overall CSS did not differ between sedatives, but there was a treatment×time interaction: IV had higher CSS at 5-20 minutes than GEL, and GEL had higher CSS at 40-150 minutes than IV ($P<0.05$). IV had lower HR than GEL during the first 10 minutes ($P<0.05$). The mean (range) administration time for LA without resistance was 11 (5-20) minutes for IV and 38 (25-55) minutes for GEL. Oro-transmucosal detomidine is an effective sedative agent for calves. Convenient and non-invasive oral sedation method for disbudding facilitates the use of local anaesthetics by cattle owners and thus, improves calf welfare.

The impact of spaying on the welfare of beef cattle

Petherick, Carol[1], Mayer, David[2], McCosker, Kieren[3] and McGowan, Michael[4], [1]The University of Queensland, QAAFI, P.O. Box 6014 Red Hill, Rockhampton QLD 4701, Australia, [2]Dept. Employment, Economic Development and Innovation, Ecosciences Precinct, Brisbane, QLD 4102, Australia, [3]Dept. Resources, Pastoral Production, Katherine Research Station, Katherine, NT 0851, Australia, [4]The University of Queensland, School of Veterinary Science, Gatton Campus, Gatton, QLD 4343, Australia; c.petherick@uq.edu.au

In parts of northern Australia, spaying can legally be conducted without anaesthesia/analgesia and is conducted via flank laparotomy (F) or per vagina (V) and F often involves electroimmobilisation (E). A comparison of the two procedures and three controls was conducted. The treatments were: F+E; V; E; mock artificial insemination (AI); and physical restraint (C), (n=20 yearling heifers and 10 cows per treatment). Welfare assessment included measures of behaviour and plasma cortisol concentrations collected at intervals to 8 h, and on days 1-4 post-procedures. Behaviours were analysed by GLM and cortisol by repeated-measures ANOVA. In the 8 h post-treatment, standing head down was shown by more ($P<0.05$) F+E cows compared to other treatments (16.5% for F+E, vs.1.6%, 1.7%, 6.8% and 2.8% for V, E, AI and C respectively, s.e.=2.02) and also for heifers (26.3% for F+E, vs. 9.8%, 7.8%, 9.0% and 3.6% for V, E, AI and C respectively, s.e.=2.13). At 8 h post-treatment, fewer ($P<0.05$) F+E and V heifers (40.1%, 56.7%, 73.6% 70.2% and 78.7% for F+E, V, E, AI and C respectively, s.e.=3.72) and cows (22.7%, 30.8%, 26.9%, 55.1% and 56.4% for F+E, V, E, AI and C respectively, s.e.=5.68) were observed feeding compared to other treatments, although F+E, V and E cows did not differ. Plasma cortisol concentrations in the 8 h post-treatment were significantly ($P<0.05$) greater in F+E (363.2 nmol/l) and V (360 nmol/l) heifers compared to C (255.6 nmol/l), with E (294.4 nmol/l) and AI (281.5 nmol/l) intermediate. F+E (299.4 nmol/l), V (296.3 nmol/L) and E (260.4 nmol/L) cows had significantly ($P<0.05$) greater cortisol concentrations compared to AI (177.5 nmol/l) and C (157.6 nmol/l). Results indicate that short-term pain and stress were similar with the two spaying methods; spaying should not be conducted without measures to ameliorate the associated pain and stress.

Infra-red thermography as a tool for quantitative pain assessment

Lomax, Sabrina, Espinoza, Crystal, Mikhaleva, Natasha and Windsor, Peter, University of Sydney, Faculty of Veterinary Science, J.L Shute Building PMB 3, Camden 2570, Australia; sabrina.lomax@sydney.edu.au

Infrared thermography (IRT) is a non-invasive technique that allows examination of skin temperature distributions by the measurement of infrared radiation emitted by an object. Tissue damage following surgical interventions results in increased blood perfusion to the site of injury as a part of the inflammatory response. This causes local heating that can be quantified by identifying changes in temperature distribution. Assessment of temperature at the site of injury may provide some information on the duration of discomfort experienced by the animal. This work aims to evaluate the feasibility of IRT to quantify inflammation following mulesing in sheep, and assess the efficacy of analgesic treatments. A randomized trial was conducted on a mob of 60 merino sheep aged 6 months undergoing routine mulesing. Sheep were allocated to 5 treatment groups to compare unmulesed controls to mulesed sheep with or without tolfenamic acid (tolf) (2 mg/kg IM) alone or in combination with topical anaesthetic (TA). Thermal photographs of the mulesing wound were taken prior to mulesing and 1 min, 5 h, 24 h, 48 h and 7 d following the procedure using FLIR E60 thermal imaging camera. Temperatures at 9 sites within and around the mulesing wound were measured and recorded using FLIR reporting software and compared in Genstat using REML to assess treatment effects. A significant treatment effect was recorded over time ($P<0.01$). The untreated and Tolf treatment groups displayed higher mean temperatures than TA and Tolf+TA treatment groups at 1 min post-mulesing ($P<0.01$). At 5 h, Tolf+TA had significantly lower mean temperatures than untreated, Tolf and TA treatment groups ($P<0.01$). After 5 h and for the remainder of the experiment, no significant changes in mean temperature were observed. IRT allows the identification of inflamed tissue, and has the potential to become a method for assessing analgesic efficacy in livestock.

State-dependent learning: a welfare assessment tool to identify the effects of different diet regimes on hunger state in broiler breeders?

Buckley, L.A.[1], Sandilands, V.[2], Hocking, P.M.[3], Tolkamp, B.J.[2] and D'Eath, R.B.[2], [1]Harper Adams University College, Animal Production, Welfare and Veterinary Sciences, Newport, Shropshire, TF10 8NB, United Kingdom, [2]Scottish Agricultural College, Kings Buildings, West Mains Road, Edinburgh, EH9 3JG, United Kingdom, [3]The University of Edinburgh, The Roslin Institute and Royal (Dick) School of Veterinary Studies, Easter Bush, Midlothian, EH25 9RG, United Kingdom; lbuckley@harper-adams.ac.uk

The use of state-dependent learning (SDL) as a novel welfare assessment tool to quantify the effect of quantitative (QFR) or qualitative dietary restriction on broiler breeder hunger was evaluated in three experiments. In each experiment, birds alternated every two days between two diet options with each diet option paired with a different coloured food reward. It was predicted that the reward associated with greatest hunger would be preferred in a subsequent choice test. Therefore, the QFR-associated reward would be preferred if the alternative diet option improved satiety. Each bird was tested twice (once per diet option fed on the test day). In experiment 1 (pilot), birds alternated between QFR and *ad libitum* (AL) access (n=4). During testing, birds preferred the QFR – associated reward during both tests (mean (± SD) proportional preference: 0.95±0.08). In experiment 2, birds alternated either between QFR and AL (QFR/AL group, n=12) or QFR and QFR + calcium propionate (100 g/kg total feed; QFR/CAP group, n=12). Only the QFR/AL birds showed a preference (preference for QFR food reward, T_{12}=12, P=0.006, test 1 only). However, differential reward intake during training by QFR/AL birds confounded the results in experiments 1 and 2. In experiment 3, birds either alternated between QFR and AL (QFR/AL group, n=8) or QFR and QFR + cellulose (100 g/kg total feed). The food reward for the latter group was based on the cellulose diet (QFR/CEL (CEL) or the QFR diet (QFR/CEL (QFR) (both n=8). All birds consumed all of the food reward on all training days. During testing, no reward preferences were found. It was concluded that any SDL-derived preferences observed were an artefact. Thus, conclusions about any satiating effects of qualitative dietary restriction were not possible.

Towards a 'depressive-like' state in horses?

Fureix, Carole[1,2], Jego, Patrick[2], Henry, Séverine[2], Lansade, Léa[3] and Hausberger, Martine[2], [1]University of Guelph (present address), Animal Behaviour and Welfare group, Animal and Poultry Science department, 50 Stone Road East, Building #70, N1G 2W1 Guelph ON, Canada, [2]University of Rennes 1, UMR CNRS 6552 Ethologie Animale et Humaine, 263 avenue du Général Leclerc, 35042 Rennes, France, [3]Université de Tours – Institut Français du Cheval et de l'Equitation, UMR 6175 INRA-CNRS Physiologie de la Reproduction et des Comportements, 37380 Nouzilly, France; carole.fureix@gmail.com

Domestic horses often encounter chronic stress, which in humans leads to a variety of negative psychological effects (e.g. depression). Pilot observations performed on 59 adult horses from three riding schools (30 min total/horse) revealed that 24% of these horses presented an atypical profile in their home environment, termed withdrawn hereafter and characterized by a particular posture (stretched neck and similar height between neck and back), the absence of ear and head movements and a fixed gaze (no visible eye movements) from 17 to 97 seconds (an apparently unusual trait, mean gaze duration of horses under natural conditions: 15±4 s). Our aim here is to investigate behavioural and physiological correlates of this atypical state, by comparing withdrawn horses (n=14) with non-withdrawn animals from the same stables (n=45). We evaluated their responsiveness to their environment (tactile reactivity, standardized sudden human approaches), their anxiety levels (novel object) and their cortisol levels (two blood samples collected for each horse between 18:00 and 19:00). Withdrawn horses appeared more indifferent to environmental stimuli in their home environment than the non-withdrawn horses (lowered responsiveness to tactile stimuli, Chi-square, $P<0.01$; indifference towards a human approach, Mann-Whitney, $P<0.01$) but reacted more emotionally in more challenging situations, i.e. novel object test (chi-square, $P<0.001$). Withdrawn horses exhibited lower plasma cortisol levels (Mann-Whitney, $P<0.05$), a phenomenon sometimes reported in various taxa when chronically stressed or depressed. Females were over-represented in withdrawn horses (chi-square, $P<0.01$), adding to the convergence with human depression. Altogether, results suggest a 'syndrome' in horses that resembles depression in humans, and open a promising line of investigation of what altered welfare states could look like in horses.

Snug as a bug in a rug: using behavioral thermoregulation to measure thermal stress and improve laboratory mouse welfare

Gaskill, Brianna N.[1], *Gordon, Christopher J.* [2], *Davis, Jerry K.* [3], *Pajor, Edmond A.* [4], *Lucas, Jeffrey R.* [3] *and Garner, Joseph P.* [5], [1]*Charles River Laboratories, 251 Ballardvale St, Wilmington, MA 01887, USA,* [2]*Environmental Protection Agency, Research Triangle Park, NC, 27709, USA,* [3]*Purdue University, 915 W State St., West Lafayette, IN 47907, USA,* [4]*University of Calgary, 3280 Hospital Drive NW, Calgary Alberta, T2N 4N1, Canada,* [5]*Standford University, 287 Campus Dr, Stanford, CA 94305, USA; brianna.gaskill@crl.com*

In these experiments, a nesting material enrichment is validated in terms of reducing a physiological stressor with natural coping behavior, improving welfare, and demonstrating end user benefits. Mice are housed at ambient temperatures (T_a) between 20-26 °C, which is below their lower critical T_a (30 °C). Thus, mice are thermally stressed, which can compromise aspects of physiology from metabolism to behavior. Raising T_a is not a solution due to increased aggressive interactions and because temperature preference differs by gender and behavior. Nest building is an adaptive behavior that increases survival and reproduction in the wild. We hypothesized that nesting material would allow mice to alleviate cold stress by building insulating nests and creating a unique microenvironment within the cage. Three common types of mice (BALB/c, C57BL/6, CD-1), of both sexes, were tested in the following experiments. Data were analyzed as GLMs. We first determined how much material (0-10 g) was preferred by mice at different T_a. Within standard $T_{a,}$ all mice preferred 6-10 g of nesting material ($P<0.05$). Thus, mice should receive no less than 6 g but may need 10 g or more to alleviate thermal stress. Second we determined if nesting material altered thermogenesis. Dome-like nests radiate less heat ($P=0.03$) and CD-1 mice ($P<0.01$) and males ($P<0.01$) showed energetic reductions from nest building. Only BALB/c mice had reduced non-shivering thermogenesis, based on brown fat protein expression ($P<0.05$). Last we tested if reduced thermogenesis from insulating nests would free resources and improve reproduction. All mice provided nesting material showed improved breeding performance ($P<0.01$) without increased food consumption ($P>0.05$). Only C57BL/6 mice had increased pup survival ($P<0.01$). This improvement has the potential to result in economic gain for producers. These experiments demonstrate that nest building decreases heat loss and energy expenditure. Reduction in thermal stress, by creating a unique microenvironment, improves mouse welfare and breeding profitability.

Detection of stress and evaluation of living conditions in any holding system by chronobiological analysis of continuously recorded behavioural data

Berger, Anne and Scheibe, Klaus-Manfred, Leibniz-Institute for Zoo and Wildlife Research, Alfred-Kowalke-Str. 17, 10315 Berlin, Germany; berger@izw-berlin.de

We offer a telemetric-chonobiological method to compare and evaluate the quality of living conditions of animals quantitatively and to follow normal changes of general state such as birth or adaptation of annual rhythms. The approach can be applied to identify stress, disturbance, disease, time of birth and endangering situations. It is especially convenient for free ranging animals but can also be applied to other conditions of animal keeping. Measuring animal behaviour continuously may contribute to investigations regarding animal-environment interactions. Alterations in living conditions result in behavioural adaptations. They can be followed by analysing level and rhythm of fundamental behaviours such as general activity measured by acceleration sensors. Time functions of accelerations from automatic telemetry systems were analyzed macroscopically (daily levels) and microscopically (autocorrelation function and power spectral analysis). Degrees of functional coupling (DFCs) were calculated to identify and evaluate disturbances in behaviour. Hierarchic frequency tuning of complex rhythmic functions of behaviours, leads primarily to period lengths that are synchronized with the 24-h period. DFCs, a measure of harmony between internal rhythms and the external 24-h period, were found to be high in well-adapted, healthy, and undisturbed individuals but were lowered during periods of adaptation, sickness, or social interaction. Thus, specific stress conditions could be identified and evaluated in several species, under various conditions, using these biorhythmic analyses. All kinds of continuous and equidistant long-term recordings of behaviour or physiology are suitable for this method. As simple behavioural parameters, such as motor activity, can be easily recorded by telemetry from animals in nearly any holding systems, it is possible to compare different holding systems. Investigations were basically carried out on horses and sheep kept under different conditions. It is the purpose of this report to describe the basic idea, and the procedure, and to give some examples of application.

When reality and subjective perception highly differ: questionnaire vs. objective observations discrepancies in horses' welfare assessment

Lesimple, Clémence and Hausberger, Martine, Université de Rennes1, UMR 6552 Laboratoire EthoS, Station Biologique, 35380 Paimpont, France; clemence.lesimple@univ-rennes1.fr

As stereotypic behaviours may reflect suboptimal life conditions, evaluating their prevalence in different environments is crucial in order to identify the problems and favour better management practices. In horses these surveys identified 1 to 10% of individuals in the tested populations performing stereotypic behaviours. However, a few observational studies converge towards much higher prevalence (22 to 96%). Different factors may explain this discrepancy, i.e. type of horse, definition of stereotypies or the methodology used. In order to disentangle these factors, we conducted a study on 373 horses of various ages and breeds, distributed amongst 26 riding schools. Questionnaires were given to caretakers, asking about the potential presence and type of stereotypies for each horse. In parallel, each horse was observed using Instantaneous Sampling method (3 sessions of 30 min). Horses were also observed continuously during 3 sessions of 6 h, to build an exhaustive list of stereotypic behaviours for each horse. The results showed a high discrepancy between the 2 types of evaluations both in the percentage of stereotypic horses detected (5.36% (Questionnaires) versus 37.53% (Observations), χ^2-test $P<0.001$) and in the type of stereotypies detected (e.g. head nodding vs. weaving, χ^2-test $P<0.001$). These results show that some stereotypies, even in 'classical' ones seem to be difficult to detect and abnormal repetitive behaviours (e.g. repetitive licking) may also not be identified as such. Thus, the under evaluation by caretakers, due to a lack of attention or recognition of stereotypies may be the primary factor in the differences observed between studies in the literature, but other aspects (unwillingness to admit the 'problem', horses having been punished when performing stereotypic behaviours, and performing less when in presence of caretakers...) remain to be tested. This study is raising the general question of using questionnaires in the assessment of behavioural traits and of their use for welfare assessment.

Qualitative behaviour assessment is independent from other parameters used in the Welfare Quality® assessment system for beef cattle

Kirchner, Marlene K.[1], Schulze Westerath-Niklaus, Heike[2,3], Gutmann, Anke[1], Pfeiffer, Christina[1], Tessitore, Elena[4], Cozzi, Giulio[4], Knierim, Ute[3] and Winckler, Christoph[1], [1]BOKU University of Natural Resources and Life Sciences, Vienna, Department of Sustainable Agricultural Systems; Division of Livestock Sciences, Gregor-Mendel-Str. 33, 1080 Wien, Austria, [2]current address: ETH Zürich, Animal Behaviour, Health and Welfare Unit; Institute of Agricultural Sciences, Universitätsstr. 2, 8092 Zurich, Switzerland, [3]University of Kassel, Department of Animal Nutrition and Animal Health, Nordbahnhofstr. 1a, 37213 Witzenhausen, Germany, [4]University of Padova, Department of Animal Science; Dairy and Beef Cattle Nutrition, Management and Welfare, Agripolis, Viale dell'Università 16, 35020 Legnaro, Italy; marlene.kirchner@boku.ac.at

The Welfare Quality® (WQ) system aims at assessing farm animal welfare using animal-based measures following a multi-criteria approach. Results of 24 single measures are aggregated to 11 'criterion' and 4 'principle' scores. Qualitative Behaviour Assessment (QBA) is used to reflect the emotional state of the animals; it is calculated using expert-derived coefficients (PC_WQ). It is debatable if QBA may be seen as a complementary measure of welfare or if it is mainly influenced by other parameters used in the assessment system. It was therefore the aim of the present study to investigate relationships between PC_WQ scores with other WQ-scores. We were also interested if results change if QBA-scores were not based on predefined coefficients but Principal Component Analysis was re-run on the data set (PC_new). 189 datasets from WQ-assessments of 63 beef farms in Austria, Germany and Italy were used for calculation of WQ-scores including PC_WQ and PC_new. Results at measure, criterion and principle level were then correlated with PC_WQ and PC_new scores. QBA-scores calculated with PC_WQ and PC_new were highly correlated ($r=0.893$, $P<0.001$). Pearson correlations of WQ results at measure level ranged from -0.27 to 0.20 for PC_WQ ($P<\pm0.05$ for 8 measures) and -0.32 to 0.23 for PC_new scores ($P<\pm0.05$ for 4 measures). At criterion level correlations ranged from -0.09 to 0.28 ($P<0.05$ for 2 criteria) and from -0.07 to 0.30 ($P<0.05$ for 3 criteria), respectively. Also at principle level (excluding the principle 'Appropriate behaviour') correlations were low (0.03 to 0.32 and 0.03 to 0.33, respectively; $P<0.05$ for 'Good housing'). We conclude that both PC_WQ and PC_new QBA-scores at any level represent information on the animal's welfare state which is independent from the other parameters included in the WQ system. There were no differences between standardized and dataset-specific QBA-scores indicating that expert-derived coefficients in WQ may be regarded as generally valid.

Assessing motivation for appetitive behaviour: food restricted broiler breeders cross a water barrier to access a foraging area without food

Dixon, Laura M.[1], Sandilands, Vicky[2], Bateson, Melissa[3], Tolkamp, Bert J.[2] and D'Eath, Rick B.[1], [1]SAC, Animal Behaviour and Welfare, SAC King's Buildings, West Mains Rd, Edinburgh EH9 3JG, United Kingdom, [2]SAC, Animal Health, SAC King's Buildings, West Mains Rd, Edinburgh EH9 3JG, United Kingdom, [3]Newcastle University, Center for Behaviour and Evolution, Henry Wellcome Building for Neuroecology, Framlington Place, Newcastle upon Tyne NE2 4HH, United Kingdom; laura.dixon@sac.ac.uk

Motivational tests for a resource have been criticised when they attempt to measure motivation while providing that resource as a reward because this itself can affect motivation during testing. Broiler breeder chickens will work hard to peck a key to access additional food, but is this because food rewards increase their motivation for food? Our previous work found that broiler breeders preferred to search areas where they had not recently been allowed access (for 2, 4 or 6 days) rather than remaining in their home pen. A new method to measure motivation for the appetitive phase of foraging without providing the consummatory resource (food) was developed: We used a runway apparatus in which birds had to pay a cost by walking through increasingly longer and deeper sections of water in order to access a foraging platform containing wood shavings but never food. Young broiler breeder hens (n=24) were fed either at commercial levels of restriction (R) or 3 times that amount (3R). Data were analyzed using mixed models in Genstat. R birds were more successful at overcoming their aversion to water and accessing the foraging platform than 3R birds (number of successful tests: R=4.92±0.43; 3R=2.83±0.58, $P<0.05$). R birds also had a significantly shorter latency to reach the foraging platform than 3R birds (R=192±65 sec, 3R=335±70 sec, $P<0.05$). This demonstrates that food restricted birds will overcome aversive stimuli for the opportunity to perform appetitive (food searching) behaviour. This work adds to the evidence that commercially restricted broiler breeders are chronically hungry which appears to be detrimental to their welfare. Finally this method relates to the 'out of sight, out of mind' problem in motivation studies: motivation for a resource can be assessed without provision of the resource itself.

Relationship between blood platelet serotonin parameters, brain serotonin turnover and behavioural responses of pigs

Ursinus, Winanda W.[1,2], Bolhuis, J. Elizabeth[2], Zonderland, Johan J.[3], Kemp, Bas[2], Korte-Bouws, Gerdien A.H.[4], Korte, S. Mechiel[4] and Van Reenen, Cornelis G.[1], [1]Wageningen UR Livestock Research, Department of Animal Sciences, Animal behaviour & Welfare, Edelhertweg 15, 8219 PH Lelystad, the Netherlands, [2]Wageningen University, Department of Animal Sciences, Adaptation Physiology Group, De Elst 1, 6708 WD Wageningen, the Netherlands, [3]Veterinary Health Research NZ, Innovation Park, Ruakura Road, Hamilton, New Zealand, [4]Utrecht University, Utrecht Institute for Pharmaceutical Sciences, Division of Pharmacology, Sorbonnelaan 16, 3584 CA Utrecht, the Netherlands; nanda.ursinus@wur.nl

Intensively kept pigs display behavioural problems that reflect maladaptation. The susceptibility to develop maladaptive mental disorders is associated with suboptimal serotonergic (5-HT) functioning in the brain, but may, as in humans, also be reflected in peripheral markers of 5-HT activity. We investigated whether blood 5-HT markers are correlated with brain 5-HT activity, and with behavioural responses. Pigs (n=32) were kept barren (according to conventional housing) or enriched (included straw, peat, wood shavings, branches and space) and exposed to a 1-min Backtest at 2 weeks. Using PCA their response (latency/frequency of struggling and vocalizing) was summarized in one component explaining 93% of variance. Blood platelet 5-HT concentration (Plt 5-HT) was measured at 13 and 19 weeks of age, and 5-HT uptake in platelets in week 13. 5-HT, 5-HIAA and 5-HT turnover (5-HIAA/5-HT ratio) levels in the Hypothalamus (H), left and right Frontal Cortex (FC) and Hippocampus (HC) were determined at 19 weeks. Pearson correlations were performed on residuals with housing as effect. Housing did not affect brain or blood 5-HT parameters. Plt 5-HT in week 13 and 19 were correlated (r=0.52). Plt 5-HT in week 13 correlated with Plt 5-HT uptake (r=0.37), with 5-HIAA in the right FC (r=0.34), 5-HT in left (r=0.46) and right (r=0.43) HC, and 5-HT turnover in left (r=0.37) and right (r=0.56) HC. Plt 5-HT in week 19 correlated with 5-HT levels in left (r=0.38) and right FC (r=0.44) and with 5-HT turnover in the left FC (r=0.41). Pigs showing an active response in the Backtest had lower Plt 5-HT in week 19 (r=-0.44). No other significant correlations were found. 5-HT in blood platelets is related to brain serotonergic parameters, and to the Backtest response in early life. Thus, peripheral serotonergic parameters may reflect brain 5-HT functioning. Further research will have to elucidate whether blood 5-HT parameters are useful markers of individual differences in adaptive capacity in pigs.

The importance of correctly classifying behavior: the case of stereotypic pacing in zoo animals

Watters, Jason, Chicago Zoological Society, Center for the Science of Animal Welfare, 3300 Golf Road, Brookfield, IL 60513, USA; jason.watters@czs.org

Most people consider stereotypic behaviours to be indicators of poor living conditions. Previous research apparently demonstrates that captive exotic animals express a high frequency of stereotypy. The husbandry conditions of most captive animals are such that food acquisition and other positive outcomes are highly scheduled. These conditions promote the development of anticipatory behaviour, yet little research has investigated whether pacing or other repetitive behaviours performed by captive animals is anticipatory rather than stereotypic. I will compare the most accepted definition of stereotypic behaviour to that of anticipatory behavior and suggest that, even though the outward expression may look the same, stereotypic behaviour and anticipatory behaviour are different. They have different causes and different consequences. The occurrence of stereotypic behaviour may offer little insight to an animal's current welfare state as it may suggest that, at some point in time, the animal experienced poor living conditions. Anticipatory behaviour on the other hand is an indicator of an animal's sensitivity to reward and as such, it is a real-time indicator of animals' own perceptions of their well-being. It is highly likely that researchers evaluating repetitive behaviours in captive animals have often confused stereotypy with anticipation. As a result, the welfare of captive exotic animals has been misestimated.

Animal welfare assessment of captive big felids in Chilean zoos using an audit system

Bravo, Camila[1], De La Fuente, Fernanda[1], Bustos, Carlos[2] and Zapata, Beatriz[1], [1]Escuela de Medicina Veterinaria, Universidad Mayor, Unidad de Etologia y Bienestar Animal, Camino la Piramide 5750, Santiago, Chile, [2]Universidad Santo Tomás, Ejercito 146, Santiago, Chile; camilona20@gmail.com

Large felids have difficulties meeting their psychological needs in captivity. However, the majority of zoos keep at least one specimen. For consistent assessment of welfare, an auditing system is needed, but standard protocols have not yet been developed for zoos. In a previous study we developed and validated a protocol to assess welfare of big felids in captivity and in this study we applied this system in four zoos in central Chile. The aim was to compare overall welfare between the zoos, to identify the main welfare problems and to compare welfare between species. The evaluation was conducted at enclosure level and measured the following principles (based on the Welfare Quality® protocol): Good Feeding (four indicators), Appropriate Housing (eight indicators), Good Health (eight indicators) and Appropriate Behaviour (13 indicators). Each indicator had a score of 0, 1 or 2, and finally a total score was assigned to each enclosure, which was transformed into an index from 0 to 1 (WI). Nineteen enclosures were assessed during 2011: four jaguar (Panthera onca), eight lion (Panthera leo), three puma (Puma concolor) and four tiger (Panthera tigris) enclosures. Four enclosures were assessed in Zoos A and B, five in C and six in D. Significant differences were found in overall WI among zoos ($F_{(1,3)}$=60.7, P<0.001), but not among species. Zoos A and C had greater WI than B and D (mean±SD: A=0.86±0.54 and 0.88±0.02 vs. 0.72±0.06 and 0.55±0.03), and D had significantly the lowest WI. The main welfare problem was in the Appropriate Behaviour principle and there were differences among zoos in this ($F_{(1,3)}$=43.66; P<0.001). The problems were related to the absence of environmental enrichment and inadequate group composition. We concluded that there are important differences in the welfare of large felids among Chilean zoos, related to their ability to express appropriate behavior.

Framework to determine a positive list for mammals suitable as companion animals

Koene, Paul, Ipema, Bert A.H., De Mol, Rudi R.M. and Hopster, Hans, Wageningen University and Research Center, Wageningen Livestock Research, De Elst 1, 6700 AH, the Netherlands; paul.koene@wur.nl

In order to estimate the suitability of animal species as companion animals, information on natural and captive behaviour, adaptation and animal welfare is essential. However, often such data are limited, not known or not formally published. In such cases, a method of predicting the welfare risks of exotic animals as potential companion animals is needed. A new framework is designed to enhance the transparency and objectivity of detecting the suitability of animal species as pet animals in an early stage. Can the animal easily adapt to the captive environment and changes in its surroundings? If an animal species finds it difficult to adapt because of its natural behavioural needs, this may lead to behavioural and welfare problems. Examples of strong behavioural needs are species with specific food requirements of specific social needs. In the first step, data on the behaviour of animal species were collected from literature and entered in a database, to assess the species behavioural needs based on statements about their natural behaviour. The possible behaviours were divided into eight functional behavioural criteria, including space, time, food, safety, maintenance of integument, reproduction, social and information requirements. The next step in the method is the assessment of the welfare risks of keeping the species in a human environment as companion animal. In the last stage, this information is combined with legal and risk factors – e.g. related to conservation, trade of endangered species, disease and danger to humans and the environment – to provide the final assessment of the suitability or potential of an animal species as a companion animal. Based on this assessment a positive list for mammals (mammals that can be legally kept) is determined. The theory and practice of determining the suitability of mammals as companion animals will be demonstrated with some striking examples of rodents.

Environmental factors that affect the behavior and welfare of domestic cats (*Felis sylvestris catus*) housed in cages

Stella, Judi[1], Croney, Candace[2] and Buffington, Tony[1], [1]The Ohio State University, Veterinary Preventative Medicine, A100P Sisson Hall 1920 Coffey Rd Columbus, OH 43210, USA, [2]Purdue University, Animal Sciences, 915 West State Street, West Lafayette IN 47907, USA; stella.7@osu.edu

The aim of this study was to examine the effects of the macro- (room) and micro- (cage) environments on cat behavior to improve the behavior and welfare of cats housed in cages. Cats (n=76) were caged singly in a vivarium at the Ohio State University and randomly assigned to one of four treatment groups: two replicates each of M+m+/M+m- and M-m+/M-m- where (M+) was a managed macro-environment (minimal noise, disruption, consistent schedule); (m+) was an enriched micro-environment (hide area, perch, consistent cage setup); (M-) was an unmanaged macro-environment (recorded dog barks, frequent disruptions, unpredictable schedule); (m-) was an unenriched micro-environment (no hide/perch area, inconsistent cage set-up). Cats were observed for 8 hrs/day for 2 days for maintenance behaviors (e.g. eating, drinking, elimination), agonistic and avoidant behaviors (e.g. growling, hissing, hiding) and affiliative behaviors (e.g. soliciting caretaker interaction) using scan sampling (9 time points) and 5 minute focal sampling (8 time points). A stranger approach test was conducted at the end of day two. Data analysis by 2-way ANOVA revealed differences in agonistic behaviors in M+ vs. M- (treatment $P<0.0001$, df=1, F=71.1; time $P<0.0001$, df=8, F=11.43; interaction P=0.003, df=8, F=1.93). m+ vs. m- revealed differences in avoidant behaviors (treatment $P<0.0001$, df=1, F=36.8; time $P<0.0001$, df=8, F=10.77; interaction P=0.6, df=8, F=0.84). Feed intake was affected only by time in M+ vs. M- ($P<0.0001$, df=1, F=40.8) and m+ vs. m- ($P<0.0001$, df=1, F=43.02). Unpaired t-tests of the effect of treatment during the approach test revealed shorter latency to interact (P=0.03), longer duration of interaction (P=0.03) and more cats approaching (P=0.008) in M+ vs. M-. These results suggest the macro-environment may be at least as relevant to the cat as the micro-environment, indicating that attention to cage enrichment without consideration for the effect of the room may be insufficient to protect captive cat welfare.

Evaluation of environmental enrichment preferences of domestic cats using a choice test

Ellis, Jacklyn J., Stryhn, Henrik, Spears, Jonathan and Cockram, Michael S., Sir James Dunn Animal Welfare Centre, Dept. of Health Management, Atlantic Veterinary College, University of Prince Edward Island, 550 University Avenue, Charlottetown, PE, C1A 4P3, Canada; jjellis@upei.ca

Preferences by cats for environmental enrichment were investigated in a choice test. It was hypothesised that cats would prefer an enriched over an empty compartment and that they would show a preference for a type of enrichment. Twenty-six cats were kept singularly in choice chambers for 10 days. Each chamber had a central area containing food, water and a litter tray, and four centrally-linked compartments containing different types of enrichment: (1) an empty control; (2) a play opportunity with a prey-simulating toy; (3) a shelf to provide a perching opportunity; and (4) a cardboard box with a hole in the side to provide a hiding opportunity. Cat movement between compartments via 'cat-flap' doors was monitored using a data-logger. Median duration/visit differed significantly between compartments (Friedman's test S=26.35, DF=3, P<0.001): 325 s (IQR=385) empty compartment, 295 s (IQR=380) toy compartment, 638 s (IQR=3545) perching compartment, and 5253 s (IQR=3779) hiding compartment. Median number of visits/day differed significantly between compartments (Friedman's test S=10.46, DF=3, P=0.015): 4 (IQR=4) empty compartment, 3 (IQR=4) toy compartment, 3 (IQR=5) perching compartment, and 5 (IQR=5) hiding compartment. Median percentage of time spent in each compartment differed significantly between compartments (Friedman's test S=24.69, DF=3, P<0.001): 3% (IQR=6) empty compartment, 2% (IQR=9) toy compartment, 6% (IQR=32) perching compartment, and 55% (IQR=46) hiding compartment. Bonferroni-adjusted sign tests showed that cats spent significantly longer/visit and spent a higher percentage of time in the hiding compartment than in the empty compartment (P<0.001) and the toy compartment (P<0.001), but was not significantly different compared with the shelf compartment. The hiding box was preferred over the absence of enrichment and the prey-simulating toys. Within the limitations of the methodology, the results suggest that a hiding box may be a resource of value to cats, and that its inclusion in enclosure design could be important.

Spatial allowances and locomotion in companion rabbits

Cooper, Jonathan, Hardiman, Jessica and Dixon, Laura, University of Lincoln, School of Life Sciences, Riseholme Park, LN2 2LG Lincoln, United Kingdom; jcooper@lincoln.ac.uk

Enclosure size may affect the locomotion in pet rabbits. There is, however, little documented evidence of the effect of enclosure size on behaviour of different sized breeds of pet rabbits. We investigated the effect of increasing floor space from median size of hutches (9,000 cm^2) to twice and four times this allowance. Rabbits were habituated to a home pen whose dimensions could be adjusted and treatments were presented systematically to control for order effects. Dwarf (<2 kg), standard (2-3 kg), large (3-4 kg) and giant breeds (>4 kg) (n=4 for each) were recorded on video over 24 hour periods in each enclosure. Data on time budgets and body postures have been presented elsewhere. We selected 4 hour-long periods (6 am, noon, 6 pm and midnight) for each rabbit for each enclosure size and recorded the number of hops, duration of locomotion, and distance travelled for all bouts of locomotion. From this data it was possible to derive measures of distance travelled and speed of movement. Data was suitable for nested ANOVA without transformation, to investigate effects of rabbit and enclosure size. Increasing floor area increased number of hops from 25±4 hops per hour in the small enclosure to 57±9 for the largest enclosure ($F=6.14$ $P<0.01$), due to increases in bouts of locomotion and number of hops per bout. Rabbits hopped for longer and further as enclosure size increased. There was no effect of size of rabbit on number of hops or time spent in locomotion, however large (1.4±0.1) and giant (1.5±0.1) rabbits moved more slowly than dwarf (2.0±0.2) and standard rabbits (1.9±0.1; $F=5.38$, $P=0.002$). These findings indicate that increasing enclosure size increased locomotion in all rabbits. Breed differences in locomotion may be due to larger rabbits being generally less active, or they may require larger enclosures to show more extensive locomotion.Evaluation of environmental enrichment preferences of domestic cats using a choice test

Effects of citronella and odourless cold air sprays on aversion in dogs

Webb, Catherine L. and Hemsworth, Paul H., Animal Welfare Science Centre, The University of Melbourne, Parkville, 3010 Victoria, Australia; c.webb@student.unimelb.edu.au

Dog training devices used to prevent problem behaviours normally incorporate aversive stimuli. To ensure learning occurs, an aversive stimulus must be timely and significant, but from a dog welfare perspective, it should not be highly aversive. This experiment examined the aversiveness of citronella and odourless spray collars, which are commonly used to prevent barking in dogs. Thirty desexed greyhounds from a university teaching colony were randomly allocated to one of three treatments, 1-s Citronella spray, 1-s Odourless cold air spray and Control. Using aversion learning, the dogs were individually introduced in 9 trials (3/day over successive days) to a location where the treatment was imposed. Variables measured included the time to move the dog to the treatment area ('transit time'), the time that pressure was applied to the dog's lead in order to move the dog ('duration of lead pressure'), heart rate before and during the trials, and saliva cortisol concentrations 10 minutes after the trials. Analysis of variance for repeated measures showed significant time and treatment effects on transit time ($P<0.001$, $P<0.05$) and duration of lead pressure ($P<0.001$, $P<0.01$), a significant time × treatment interaction on duration of lead pressure ($P<0.01$) and a strong tendency for time × treatment effect on transit time ($P=0.056$). The Citronella and Odourless treated dogs showed a significantly ($P<0.05$) longer transit time and duration of lead pressure than Control dogs in trial 9. There were no significant ($P>0.05$) treatment effects on heart rate or cortisol concentration. This experiment indicates that while dogs showed aversion to the sprays, these stimuli did not evoke a detectable physiological stress response in dogs. Thus based on aversion, the use of citronella and cold, odourless spray collars in dog training are unlikely to pose a significant welfare risk.

The effect of straw length on penmate directed behaviour in pens with growing pigs

Steinmetz, Henriette, Danish Agriculture & Food Council, Pig Research Centre, Axeltorv 3, 1609 Copenhagen, Denmark; hst@lf.dk

The aim of this study was to investigate whether chopped straw resulted in the same level of pen-mate directed behaviour (PB) among finishers on a Danish commercial pig farm as observed among pigs given full-length straw corresponding to 100 g straw/pig/day. The experimental subjects were 39 finisher pens each holding 15 assigned to one of two treatments: full-length straw or chopped straw. Once a day, pigs were provided with new straw corresponding to 100 g straw/pig/day. In each pen, 3 focal animals were randomly selected. Focal animals were balanced according to size and gender. A total of 117 focal animals were selected. Each pen was video recorded when the pigs weighed 40 kg and 80 kg. With PB as the primary parameter, the behaviour of the focal animals were recorded for 15 minutes per hour between 6 am and 11 pm. The duration of PB was analysed with a mixed linear model using PROC MIXED in SAS. Straw length and weight (age) were included as systematic effects, and batch and pen as random effects. The level of PB was analysed in proportion to active time. PB was expressed as an average of the 3 focal animals per pen/day. The pigs were significantly more active at 40 kg than at 80 kg (mean pct. active time; 29 versus 22 for 40 kg and 80 kg respectively giving full length straw, SD 5.5, $P<0.0001$). Straw length did not significantly affect pen-mate directed behaviour (mean pct. PB of active time; 6.2 versus 6.9 for full-length straw and chopped straw respectively at 40 kg, SD 4.2). In conclusion, no differences were observed between full-length and chopped straw on the level of pen-mate directed behaviour when straw was supplied once daily corresponding to 100 g/pig/day.

On the impact of structural and cognitive enrichment on learning performance, behaviour and physiology of dwarf goats (*Capra hircus*)

Meyer, Susann and Langbein, Jan, Leibniz Institute for Farm Animal Biology, Behavioural Physiology, Wilhelm-Stahl-Allee 2, 18196 Dummerstorf, Germany; s.meyer@fbn-dummerstorf.de

Intensive husbandry only partially allows farm animals to perform species-specific behaviour and means little challenge to their cognitive capabilities. While the positive impact of structural enrichment on animal welfare has been evaluated in a number of studies, little is known about cognitive enrichment. Utilizing a 2×2 design, we investigated the effect of structural-enriched housing (litter, round feeder, climbing rack) on learning performance and of structural and cognitive enrichment on behavioural and physiological reactions of juvenile dwarf goats. Two experimental groups (structural enriched or barren housing) got drinking water for free, while two further groups (enriched or barren) got water only after continuously solving visual four-choice discrimination tasks intended for cognitive enrichment (group size 8-9; pen size 12 m^2). The latter two groups were trained consecutively on three different tasks each consisted of four different black 2D-symbols presented on a computer screen and ran for 14 days. Before the first and after each task, all animals were individually tested for 10 min in an open-field/novel-object test to assess differences in behaviour. In addition, saliva was collected from all animals 1 h before and directly after each test to determine cortisol concentration. Structural enrichment had an increasing impact on learning performance in the course of the three tasks. Animals from enriched environments learned the third task faster than barren housed goats ($F_{1,15}$=4.54; P=0.05). In the open-field test, structural enrichment×test repetition positively affected activity ($F_{3,93}$=2.34; P<0.1) and number of rearings ($F_{3,93}$=3.55; P<0.05). Different parameters regarding contact to the novel object were positively influenced by cognitive enrichment (all P<0.05). Saliva cortisol was affected by the interaction between time of sampling×cognitive enrichment ($F_{1,213}$=3.42; P<0.1). In conclusion, structural enrichment of housing conditions can enhance the learning performance of dwarf goats. Furthermore, it seems that structural and cognitive enrichment promotes different aspects of its behavioural competence in challenging situations.

Multilevel caging enhances the welfare of rats as assessed by a spatial cognitive bias assay

Hickman, Debra and Swan, Melissa, Indiana University, Laboratory Animal Resource Center, 975 W Walnut St (IB008), Indianapolis, IN 46202, USA; hickmand@iupui.edu

This IACUC approved study was designed to evaluate the welfare of rats housed in a commercially available multilevel caging system. Thirty-six male Sprague Dawley rats were randomly assigned to one of the following groups: (1) limit to bottom level, no change; (2) started with both levels, no change; (3) limit to bottom level, change to both levels; or (4) started with both levels, limit to bottom level. They were trained to perform a spatial discrimination task; the housing environment was changed (per treatment group); and they were subsequently tested with three ambiguous 'probe' locations (reward plus, neutral, or non-reward plus). At the end of testing, a 50 µl saphenous blood sample was collected to measure immunological markers of distress. The latency times for all locations were compared between groups (ANOVA). The following findings were consistent across probe locations, but only average latency data for the neutral location is presented here. The group 3 average latency time was significantly decreased as compared to the other groups (12.7±6.1 sec, P=0.0148), suggesting a positive cognitive bias when provided with enhanced caging access. The group 4 average latency time was significantly increased as compared to the other groups (36.8±5.6 sec, P=0.0271), suggesting a negative cognitive bias when enhanced caging access was removed. There were no differences in the average latency times for neutral locations in groups 1 and 2, where there were no changes in their environment (29.1±5.8 sec and 24.3±6.1 sec, respectively, P=0.5039), suggesting habituation. However, group 2 had a significant reduction in the neutrophil:lymphocyte ratio as compared to group 1, suggesting decreased stress (0.20±0.1 and 0.37±0.1, respectively, P=0.0002). These findings suggest the welfare of these rats was enhanced through access to both levels of multilevel caging.

Preference for structured environment in zebrafish (*Danio rerio*) and checker barbs (*Puntius oligolepis*)

Kistler, Claudia[1], Hegglin, Daniel[2], Würbel, Hanno[3] and König, Barbara[1], [1]Institute of Evolutionary Biology and Environmental Studies, University of Zurich, Winterthurerstrasse 190, 8057 Zurich, Switzerland, [2]SWILD, Urban Ecology & Wildlife Research, Wuhrstrasse 12, 8003 Zurich, Switzerland, [3]Veterinary Public Health & Animal Welfare, University of Bern, Schwarzenburgstrasse 155, 3003 Bern, Switzerland; claudia.kistler@ieu.uzh.ch

In this study, we investigated the influence of environmental complexity on compartment preference and behaviour in zebrafish (*Danio rerio*) and checker barbs (*Puntius oligolepis*). For the preference test, aquaria were divided by walls of Plexiglas into an empty compartment, a structured compartment enriched with plants and clay pots, and a smaller compartment in-between. The empty and structured compartments were divided into six zones of similar size by defining three vertical layers and two horizontal areas (back vs. front area). Seven groups of six to nine zebrafish and seven groups of seven or eight checker barbs were observed on four days each. To assess compartment use, the Jacobs' preference index (JPI) was calculated and tested using a one-sample t-test. To quantify behavioural diversity, the Shannon index of diversity, and to quantify space use, the spread of participation index (SPI) was calculated. Percentages of socio-negative, socio-positive and courtship behaviour were calculated. Wilcoxon matched-pairs signed-rank tests were used to determine the differences between compartments. For the analysis of behavioural data, one group of each species had to be excluded due to constraints in daily observation time. Both zebrafish and checker barbs showed a significant preference for the structured compartment (JPI: t=3.41, df=6, P=0.01; t=9.56, df=6, P=0.0). In neither species did behavioural diversity differ between the empty and structured compartment. In the structured compartment, checker barbs showed significantly more socio-negative behaviour (Z=−2.201, P=0.028, n=6). Checker barbs used the structured compartment more evenly than the empty compartment ($SPI_{structured}$=0.46, SPI_{empty}=0.62; Z=-1.992, P=0.046, n=6). These results suggest that zebrafish and checker barbs have a preference for complex environments, and that behavioural and ecological needs may vary depending on species.

The 'UMB farrowing pen' – a promising system for future loose housing of lactating sows

Andersen, Inger L.[1], Cronin, Greg[2], Trøen, Cathinka[1], Ocepek, Marko[3] and Bøe, Knut E.[1], [1]University of Life Sciences, Animal and Aquacultural Sciences, P.O. Box 5003, 1432 Ås, Norway, [2]University of Sydney, Faculty of Veterinary Science, 425 Werromby road, Camden, NSW, Australia, [3]Univerza v Mariboru, Fakulteta za kmetijstvo in biosistemske vede, Katedra za živinorejo, Pivola 10, SI-2311 Hoče, Slovenia; inger-lise.andersen@umb.no

The aim of the present project is to provide production results and behavioural data from a new farrowing pen tested on a small scale on Norwegian and Australian sows. The 'UMB farrowing pen' (7.9 m^2) comprises two compartments: a 'nest area' and an activity/dunging area with a threshold in between. The nest area has solid side walls, sloped walls on three of the sides and a hay rack on the fourth wall allowing free access to hay or straw. The nest area has two zones with floor heating covered by a 30 mm thick rubber mattress. 12 Norwegian and 28 Australian sows were housed in the UMB pen from one week before farrowing until weaning. The effects of parity and litter size were analysed using a Genmod procedure in SAS. Mortality of live born piglets was 12.1±2.9% in the Norwegian sows and 12.9±2.0% in the Australian sows whereas the number of weaned piglets was 12.1±0.4 and 9.1±0.3, respectively. Overlying and starvation were the most common causes of death and were significantly affected by parity ($P<0.05$) and litter size ($P<0.05$). All sows showed a high level of communication with their piglets, and primiparous sows communicated significantly more with their newborns during the birth process than the pluriparous sows ($P<0.05$). At parturition, 33.3% of the sows were resting with the back towards the back wall whereas 41.7% rested towards the threshold. In 50% of the nursings, the sows were resting against the back wall while 30% of the nursings occurred towards the threshold. The preliminary production results in both countries were better than thos reported in a national data base from more than 400 commercial pig farms with individually loose-housed sows, indicating that this might be a promising farrowing pen for future loose-housing of lactating sows.

Effects of social isolation and environmental enrichment on laboratory housed pigs

Deboer, Shelly P.[1,2], Garner, Joseph P.[2], Eicher, Susan D.[1], Lay, Jr., Donald C.[1] and Marchant-Forde, Jeremy N.[1], [1]USDA-ARS, LBRU, 125 S. Russell St., West Lafayette, IN 47907, USA, [2]Purdue University, Department of Animal Sciences, Lilly Hall, West Lafayette IN 47907, USA; marchant@purdue.edu

The pig is an increasingly important laboratory animal species. However, a laboratory often requires individual, sterile housing, which may impose stress. Our objective was to determine the effects of isolation and enrichment in pigs housed for 7 days within the PigTurn™ – a novel penning system with automated blood sampling. Sixteen castrated male (7) and female (9) (Yorkshire × Landrace) weaner pigs were randomly assigned to one of four treatments in a 2×2 factorial combination of enrichment (non-enriched or enriched) and isolation (visually isolated or able to see another PigTurn). Pigs underwent catheterization and were placed into the PigTurns 48 h post-recovery. They were fed each morning and had *ad libitum* access to water. Blood was collected automatically daily at 09:00 and 13:00. The morning sample was assayed for cortisol and TNF-α, whereas the afternoon sample was used to determine WBC differentials. Behavior was video-recorded and sampled using instantaneous 10-minute scans, to determine behavior and posture time budgets. Data were analyzed as a REML using the MIXED procedure of SAS. Behavior and TNF-α was not affected by treatment. Compared to non-enriched pens, enrichment tended to increase time standing (0.31±0.02 vs. 0.34±0.03, $P<0.1$) and lying laterally (0.27±0.02 vs. 0.34±0.03, $P<0.1$) and decrease plasma cortisol (1.8±0.2 μg/dl vs. 1.3±0.1 μg/dl, $P<0.05$). There was a significant isolation × enrichment interaction. Enrichment given to pigs housed in isolation had no effect on plasma cortisol (1.5±0.2 and 1.5±0.2 μg/dl), but greatly reduced it in non-isolated pigs (1.1±0.2 and 2.2±0.2 μg/dl). Eosinophil count differed between treatments, being highest in the enriched/not isolated treatment (0.6±0.1 Kμl) and lowest in the not enriched/isolated treatment (0.2±0.1 Kμl, $P<0.05$), with the other treatments intermediate. The results suggest that being able to see another animal but not interact may be frustrating. The combination of no enrichment and isolation maximally impacts eosinophil numbers. Appropriate enrichment coupled with proximity of another pig would appear to improve welfare.

Optimising nest design for the PigSAFE free farrowing pen

Edwards, Sandra[1], Seddon, Yolande[1], Rogusz, Alicja[1], Brett, Mark[1], Ross, David[2] and Baxter, Emma[2], [1]Newcastle University, School of Agriculture, Food and Rural Development, Agriculture Building, Newcastle upon Tyne NE1 7RU, United Kingdom, [2]SAC, Animal and Veterinary Sciences, West Mains Road, Edinburgh EH9 3JG, United Kingdom; sandra.edwards@ncl.ac.uk

The PigSAFE (Piglet and Sow Alternative Farrowing Environment) system was designed to provide higher welfare housing during farrowing and lactation, incorporating features to meet biological needs of animals and practical needs of staff. Twelve farrowing pens were constructed to the same basic PigSAFE design; a nest area with straw and piglet protection features, heated creep, slatted dunging area and lockable sow feeder, in a total pen area of 7.7 m^2. Two nest designs, and two levels of straw provision were tested in a 2×2 factorial design with 25 litter replicates. Design treatment compared pens with noise suppression material lining nest walls and a sound insulated roof (QUIET) against controls (CONT). Substrate treatment compared a minimum amount of substrate (MIN; 2 kg of long straw provided at entry on day -5 and replenished to maintain this level until day +2 post-farrowing), with a more generous straw allowance (MAX; 4 kg). Continuous video recordings were made 16 h prior to first birth (BFP) until 24 h post-farrowing and standard production records taken. Data were analysed by two-way ANOVA using GLM procedure (Minitab 15), with covariates applied where appropriate. Neither design nor substrate treatment had a significant influence on duration of nest-building behaviour prior to BFP (e.g. straw directed behaviour: MAX=123, MIN=101, sem 14.3 min, P=0.27), farrowing location, farrowing duration, frequency of sow posture changes (e.g. BFP to 24 h MAX=188, MIN=155, sem 16, P=0.14; QUIET=180, CONT=162, P=0.42), maternal responsiveness in a piglet scream test (e.g. latency to respond: QUIET=5.4, CONT=5.1, sem 0.56 s, P=0.72), or piglet survival. In the absence of such differences, the management advantages of an open pen for better visibility and ease of working, and of lower substrate level for better manure handling, suggest these to be acceptable commercial recommendations. The PigSAFE system is a promising alternative to crates, but requires wider commercial testing.

Behaviour response to two-step weaning is diminished in beef calves previously submitted to temporary weaning with nose flaps

Hötzel, Maria J.[1], Quintans, Graciela[2] and Ungerfeld, Rodolfo[3], [1]Universidade Federal de Santa Catarina, Laboratório de Etologia Aplicada e Bem-Estar Animal, Depart. de Zootecnia e Desenvolvimento Rural, Rod Admar Gonzaga 1346, 88034-001, Brazil, [2]Instituto Nacional de Investigación Agropecuaria, Treinta y Tres, Uruguay, [3]Universidad de la República, Departamento de Fisiología, Facultad de Veterinaria, Montevideo, Uruguay; mjhotzel@cca.ufsc.br

Temporary weaning (TW) of beef calves for 5-14 days is a husbandry practice that is frequently used in South America to help advance rebreeding in postpartum cows. The aim of this study was to compare the behaviour response to two-step weaning with the use of nose flaps, in beef calves that had or not been submitted to TW for rebreeding of their dams. Thirteen calves that had been fitted with a nose-flap anti-suckling device from 71 to 84 days of age (TW group), and 11 calves that had not received nose flaps during rebreeding (C group) were weaned with a two-step procedure. To this end, at 189 days of age all calves were fitted with nose flaps and remained with the dams; 14 days later they were separated from the dams. Behaviour was observed from 186 to 193 days of age (nose-flap period, NF) and from 200 to 209 days of age (permanent separation period, PS). Behaviour data were analysed with mixed models for repeated measures. Before weaning, TW calves had greater suckling time ($F=6.5$; $df=1$; $P=0.018$), grazed more ($F=10.2$; $df=1$; $P=0.004$), played less ($F=5$; $df=5$; $P=0.0003$), and remained closer to the dam compared to C calves. During NF, C calves had lower frequencies of standing ($F=9$; $df=5$; $P<0.0001$) and grazing ($F=4.84$; $df=1$; $P=0.038$) indicating that calves that had nose flaps during their dams' breeding period adapted faster to the anti-suckling device and the cessation of suckling. During PS, C calves grazed less ($F=15.4$; $df=5$; $P<0.0001$) and vocalized more ($F=2.97$; $df=5$; $P<0.015$); otherwise, TW calves displayed an overall similar, but earlier and shorter behaviour response than C calves. We conclude that the benefits of step-weaning with nose flaps are greater for calves that have worn nose flaps for temporary weaning during the breeding of their dams.

Can pasture access contribute to reduced agonistic interactions and relaxation in the loose housing barn in horned dairy cows?

Irrgang, Nora and Knierim, Ute, Farm Animal Behaviour and Husbandry Section, University of Kassel, Nordbahnhofstrasse 1a, 37213 Witzenhausen, Germany; irrgang@uni-kassel.de

It can be expected that conditions on pasture, i.e. general space allowance, surface quality and low competition for feeding and lying places lead to increased relaxation in cows while they are on pasture. However, it was the aim of this study to investigate whether an effect can still be detected when the cows are back in their housing environment. Four groups of each 18-19 horned dairy cows had in succession no (p0), four (p4) and eight hours (p8) daytime pasture access for one week, respectively. Heart rate, heart rate variability, latency to lie down and lying time during 2 hours after afternoon feeding in the barn as well as agonistic interactions after morning feeding in the barn were recorded in 6 low ranking focal cows per group. Focal cows showed significantly lower heart rates during p8 (74.13 bpm) than p4 (80.67 bpm) or p0 (80.66 bpm; analysis of variance with repeated measures; F=20.556, P<0.001; post hoc: LSD), and a higher SDNN during p4 (20.36 ms) than p0 (16.00 ms, p8: 17.14 ms; F=3.662, P=0.035). Latency to lie down was shorter during p4 (44 min) and p8 (51 min) than p0 (97 min; F=24.947, P<0.001), lying times were longest during p4 (60%), followed by p8 (48%) and then p0 (32%; F=30.152, P<0.001), Agonistic interactions/30 min were less frequent during p8 (3.2) and p4 (4.2) than p0 (8.5; F=23.724, P<0.001). Highest indications of stress level were always found with no pasture access, but different access times did not always produce consistent results, possibly due to further factors such as weather conditions. However, in general, relaxing effects of pasture access could be detected in the cows in the barn. Especially the reduction of agonistic interactions is important in horned cows.

Lying behaviour of primi- and multiparous cows in early lactation

Dippel, Sabine[1], Schröder, Jennifer[1], Gutmann, Anke[2], Winckler, Christoph[2] and Spinka, Marek[3], [1]Friedrich-Loeffler-Institut, Institute of Animal Welfare and Animal Husbandry, Doernbergstrasse 25/27, 29223 Celle, Germany, [2]University of Natural Resources and Applied Life Sciences (BOKU), Division of Livestock Sciences/Department of Sustainable Agricultural Systems, Gregor-Mendel-Str. 33, 1180 Wien, Austria, [3]Institute of Animal Science, Přátelství 815, 104 00 Praha Uhříněves, Czech Republic; sabine.dippel@fli.bund.de

While lying down is an important aspect of recovery after calving, primiparous cows may have problems to do so due to their usually lower social rank. Thus, our aim was to describe the lying behaviour of primiparous (PP) and multiparous (MP) cows during their first three weeks in the milking herd after calving. Data were collected in a 52-cow post-calving group on a 260-cow dairy farm. Stocking density was 0.9-1.2 cows per cubicle. Lying duration and number of bouts were recorded with data loggers attached to a hind leg. The influence of parity and day spent in group (DIG) on lying behaviour was analysed with linear mixed models for repeated observations. N ranged from 50 cows on DIG 1 (20 PP, 30 MP) to 40 cows on DIG 21 (16 PP, 24 MP). Daily lying time increased with DIG for both PP and MP cows. While MP cows lay down longer than PP cows after entering the group (DIG 1: 10.7 vs. 8.4 h/24 h; $P<0.05$), increase in lying time was higher in PP cows (MP lying time at DIG 7, 14 and 21: 11.0, 11.3 and 11.6 h/24 h; PP: 8.9, 9.5 and 11.1 h/24 h; $P<0.05$). Number of lying bouts increased from 9.9 (SD 3.5) on DIG 1 to 12.1 bouts per 24 h (SD 6.3) on DIG 21. Parity had no significant effect on daily number of lying bouts, yet this might be due to high variation in PP cow data and will be revised after analysis of the full dataset. In sum, dairy cow lying behaviour changes during the first three weeks after calving. Multiparous cows lie down longer than primiparous cows throughout this period, while primiparous cows are increasing their daily lying time. This indicates that primiparous cows need longer to adapt to early lactation surroundings than multiparous cows.

Effect of uncomfortable standing surfaces on restless behavior and muscle activity of dairy cows

Rajapaksha, Eranda and Tucker, Cassandra B., University of California, Davis, Animal Science, One Shields Avenue, 95616, Davis, CA, USA; earajapaksha@ucdavis.edu

In dairy cows, both restless behavior and variation in weight distribution increase with lameness, but do not differ between surfaces that vary in comfort (e.g. compressibility). Our objective was to combine muscle activity with restless behavior to better understand these types of discomfort. We compared standard concrete under all 4 hooves (STANDARD) to an uncomfortable surface under all 4 hooves (2×2 cm trapezoidal-prism-shape concrete protrusions 4cm above the surface, ROUGH) and, in order to represent lameness, this same surface under a single hind hoof, while the other 3 legs were on standard concrete (MIXED). Twenty-four healthy Holstein cows were subjected to each treatment for 1 hour/d in a repeated-measures design. Electromyograms (EMG) were used to evaluate both total muscle activity and the number of changes in muscle activity (shifts) from one hind leg to the other during standing. EMG (2 muscles/hind leg; 4 total) and restless behavior, (number of steps for all 4 legs), were compared with general linear models. Cows took more steps with the hind leg on the rough quadrant of MIXED compared to the same leg on other surfaces (MIXED:206, ROUGH:87, STANDARD:86 steps, SE:14.4, $P<0.001$), indicating a more frequent stepping when discomfort is only under 1 hoof. There was more total muscle activity on the contralateral leg in the MIXED treatment (5.42, ROUGH:3.46, STANDARD:3.48 mVs, SE:0.247, $P<0.001$); muscle activity was also more sustained in this leg, as it had fewer shifts away from it (MIXED:7, ROUGH:18, STANDARD:19 shifts,SE:1.2, $P<0.001$).This suggests that cows were compensating for discomfort under 1 hoof by using muscles in the unaffected hind leg. There were no differences between the STANDARD and ROUGH treatments in any variable. Restless behavior and EMGs detect discomfort under a single leg, but are not useful in detecting discomfort under all 4 hooves.

Cortisol in faeces of fattening bulls: effect of housing conditions, temperature, and behavior

Zerbe, Frank and Petow, Stefanie, Institute of Animal Welfare and Animal Husbandry, Friedrich-Loeffler-Institute, Doernberstrasse 25/27, 29223 Celle, Germany; frank.zerbe@fli.bund.de

Cortisol is an indicator of distress but is also known to be higher in animals which behave more active. In this study the effects of floor type, stocking density (SD), temperature, and behavior on faecal cortisol were tested in cross breed bulls. In 4 batches a total of 224 bulls were kept in groups of 7 either on concrete (CON) slatted floor at SD of 2.5 and 3.0 m^2/bull or on rubber (RUB) topped slatted floor at SD of 2.5, 3.0, 3.5, 4.0, 4.5 and 5.0 m^2/bull. At a weight of 450 kg and 600 kg individual behavior was video-recorded over 72 h (number and duration of lying periods, number of mounting). Directly after observations rectal derived feaces were sampled at 10 to 12 a.m. (A) and 2 to 4 p.m. (B) on 2 days /week for 4 weeks. Based on mean outdoor temperature of the 8 sampling days temperature classes were defined: <8 °C, 8-18 °C, and >18 °C. Cortisol was measured using a saliva ELISA kit validated before. Linear mixed model (SAS 9.2 GLIMMIX) was used to analyze the effect of sampling time, weight period, temperature, floor, SD and of the 3 behavioral variables on cortisol, mean differences were tested by TUKEY test ($P<0.05$). Sampling time ($F_{1;40}=13.71$, $P<0.0006$), floor × temperature ($F_{2;33}=7.15$, $P<0.0026$) and mounting ($F_{1;33}=4.65$, $P<0.0384$) significantly affected cortisol concentrations whereas lying did not. Mean cortisol (±se in ng/g) was higher at 8-18 °C on RUB compared to CON for A (CON 12.9±1.4; RUB 21.9±1.0) and B probes (CON 13.5±1.2; RUB 27.0±0.9). In addition, values in RUB pens were also higher at 8-18 °C compared to <8 °C and >18 °C. The effect of floor type on cortisol at 8-18 °C is likely to reflect the increased activity on RUB rather than a stress response. The significant effect of the interaction floor × temperature demands to consider environmental effects when interpreting physiological data.

Thermoregulatory behaviour and use of outdoor yards in dairy goats

Bøe, Knut E. and Ehrlenbruch, Rebecca, Norwegian University of Life Sciences, Department of Animal and Aquacultural Sciences, P.O. Box 5003, 1432 Aas, Norway; knut.boe@umb.no

The aim of this experiment was to investigate thermoregulatory behaviour in dairy goats and how they use an outdoor yard under different weather conditions. The experiment was performed in an open building with four pens (1.80 m^2 per goat). Half of the total area was a resting area inside the building on deep straw bedding, and the other half was an outdoor yard with concrete surface. A 2×2 factorial experiment was conducted with roof covering of outdoor yard (yes or no) and location of feed (indoors or in yard). Four groups of five pregnant dairy goats were used in the experiment and the groups were systematically rotated between pens every week. The goats were video recorded for two 24 h periods at the end of each experimental period. Average air temperatures were ranging from -15 °C to +10 °C. When the air temperature dropped and when there was rain or snow, the goats spent less time in the outdoor yard ($F=7.3$, $P<0.0001$), but total resting time and time spent feeding was not affected by the weather conditions ($F=0.9$, $P>0.10$). Irrespective of weather conditions, the goats spent significantly more time in the outdoor yard in pens when the outdoor yards were covered with a roof (25.2 vs. 22.2% of total observations, $F=6.4$, $P<0.05$), but time spent resting was not affected by roof cover nor feed location 53.3 vs. 55.5% of total observations, $F=2.3$, $P>0.10$). Time spent resting alone decreased ($F=3.2$, $P<0.05$) and time spent resting in body contact with one other goat increased ($F=3.1$, $P<0.05$) when the air temperature decreased. We conclude that even if the outdoor yard was less used at decreasing temperatures, the time spent resting and feeding was not affected by inclement weather.

Food colour preferences in immature red junglefowl and broiler chickens

Hothersall, Becky, Caplen, Gina, Rose, Katie, Nicol, Christine J. and Knowles, Toby, University of Bristol, School of Veterinary Science, Langford, North Somerset, BS40 5DU, United Kingdom; b.hothersall@bristol.ac.uk

Differences in neophobia have been found between red junglefowl (*Gallus gallus*) and laying hens. Comparisons with broiler (meat) chicken breeds have not been made. We investigated feeding behaviour in 20 immature red junglefowl and 20 five-week old Flex Hubbard broilers under artificial tungsten lighting. Birds were simultaneously offered familiar (uncoloured) commercial feed and foods dyed with red, blue, yellow or green food colouring for one hour per day across five days. Using the multilevel modelling software MLwiN v2.22, we modelled for each colour the mass of food eaten and individuals' total frequency and duration of pecking bouts (from video footage) on Days 1, 3 and 5. In junglefowl, frequency of pecking bouts decreased across days ($P<0.001$). Uncoloured food attracted more ($P<0.001$) pecking bouts and a greater mass of uncoloured food was consumed than any other colour ($P<0.001$). Duration of pecking at uncoloured food was also higher than for any other colour and remained constant across days whereas it decreased for all other colours ($P=0.003$). Broilers showed no clear overall colour preference. Instead, significant interactions were observed between colour and day for frequency ($P=0.005$) and duration ($P=0.001$) of pecking bouts and for mass eaten ($P=0.046$). For all three measures, this was explained by an increase in red across days, with complementary declines for uncoloured and yellow food and little or no change for blue or green food. Junglefowl showed strong, consistent preference for familiar, uncoloured food. The contrasting lack of overall colour preference in broilers suggests a reduction in neophobia, probably as a product of selection for rapid weight gain. Their rapid development of a preference for red may relate to visibility or experience of red bell drinkers in the home pens. The role of taste and further persistence of colour preferences remain to be explored.

Weaning dairy calves off milk according to their ability to eat solid feed reduces the negative effects of weaning on energy intake and weight gain

De Passillé, Anne M. and Rushen, Jeffrey, Agriculture and Agri-Food Canada, 6947 Highway 7 Agassiz, V2P 7P9 BC, Canada; annemarie.depassille@agr.gc.ca

Dairy calves can show reduced growth, lowered energy intake and increased hunger when weaned off milk. Calves differ in their ability to eat solid feed, and automated feeders allow weaning to be adjusted to an individual calf's intake. We examined effects of weaning according to solid feed intake on age at weaning, energy intake, and behavioural signs of hunger. 40 group-housed Holstein calves fed from automated feeders were gradually weaned when voluntary intake of grain starter reached a target level with weaning complete when intakes were either 1,600 g/d (HIGH-END) or 800 g/d (LOW-END). Digestible energy (DE) intakes were calculated from milk, starter and hay intakes, calves were weighed and visits to the milk feeders were measured daily. Treatment effects were tested with GLM. There were large differences between calves in the age at which weaning was complete (range 43-81 d), with weaning completed earlier for LOW-END calves (59.6±2.3 d) compared to HIGH-END calves (66.8±2.2 d) ($P<0.05$). There were no differences ($P>0.10$) between the weeks before and during weaning (from when milk allowance was first reduced to when the calves received no milk) in DE intakes (0.09±0.01 vs. 0.08±0.01 Mcal/kg BW) or weight gains (1.11±0.09 vs. 1.16±0.08% body weight) and no effect of weaning method ($P>0.10$). Daily visits to the milk feeder did increase during weaning (before vs. during weaning: 8.70±0.56 vs. 15.78±1.18 /d). Neither DE intakes nor visits to the milk feeder were affected by weaning method ($P>0.10$). During weaning, the daily weight gain was higher for HIGH-END calves compared to LOW-END calves (1.42±0.10 vs. 0.88±0.11% body weight; $P<0.01$). Weaning individual calves according to their ability to eat solid feed reduces the impact of weaning. Weight gain during weaning is highest when complete weaning is delayed until calves are eating a larger amount of solid feed.

Cross sucking in milk fed calves may be motivated by a need for oral stimulation and develop into a habit over time

Vaughan, Alison[1], Miguel-Pacheco, Giuliana G.[2], Rushen, Jeffrey[3] and De Passillé, Anne M.[3], [1]University of Saskatchewan, Department of Large Animal Clinical Sciences, Western College of Veterinary Medicine, University of Saskatchewan, Saskatoon, S7N 5A8, Canada, [2]University of Nottingham, School of Veterinary Medicine and Science, Sutton Bonington Campus, Loughborough, LE12 5RD, United Kingdom, [3]Pacific Agri-Food Research Centre, Agriculture and Agri-Food Canada, P.O. Box 1000, 6947 Highway 7, Agassiz, BC, V0M 1A0, Canada; Alison.Vaughan@usask.ca

Cross-sucking among dairy calves is a concern in older animals where it may cause health problems for the recipient. Although the majority of calves stop cross-sucking following weaning, some calves continue this behaviour. Both hunger and the taste of milk increase sucking motivation. Our aim was to identify what motivates cross-sucking in high milk fed calves and to describe changes to this behaviour over time. 55 Holstein calves were housed in groups of 4-8 and fed with automated feeders. Daily duration of cross-sucking, feed intake and frequency of unrewarded visits to the milk feeder (a behavioural sign of hunger) were recorded for individual animals before and after weaning off milk. Mixed linear regression models, with group as a random effect, were used to determine factors influencing cross-sucking at each observation period. Before weaning the majority of cross-sucking events (75%) occurred >30 minutes after a milk feeder visit and thus appeared not to be motivated by the taste of milk. Increased digestible energy and decreased milk intake both increased time spent cross-sucking (P=0.02 and P=0.04, respectively). Immediately after weaning, cross-sucking was associated with a higher intake of hay (P=0.01). This may suggest that calves with high motivation for oral manipulation adopt different strategies (either increasing hay intake or sucking on the empty teat) when no longer able to suck milk through teats. Although calves preferentially cross-sucked certain individuals, fixed reciprocal cross-sucking pairs only became established after weaning. A tendency for a positive correlation between cross-sucking and being cross-sucked found before weaning (r_s=0.26, P=0.05) became highly significant after weaning (r_s=0.53, P<0.001). In this study cross-sucking may be motivated by a need for oral stimulation rather than milk ingestion or hunger and develop into a habit over time.

What do calves choose to eat and how do preferences affect calf behaviour and welfare?

Webb, Laura[1], Van Reenen, Kees[2], Engel, Bas[3], Berends, Harma[4], Gerrits, Walter[4], De Boer, Imke[1] and Bokkers, Eddie[1], [1]Animal Production Systems, Wageningen University, P.O. Box 338, 6700 AH Wageningen, the Netherlands, [2]Livestock Research, Wageningen University and Research Centre, P.O. Box 65, 8200 AB Lelystad, the Netherlands, [3]Biometris, Wageningen University, P.O. Box 100, 6700 AC Wageningen, the Netherlands, [4]Animal Nutrition, Wageningen Univeristy, P.O. Box 338, 6700 AH Wageningen, the Netherlands; laura.webb@wur.nl

Veal calves develop abnormal oral behaviours (AOB) due to lack of ruminating. We investigated what diet calves selected in a free choice situation and how this affected behaviour. Group-housed bull calves were either given unlimited access to milk replacer (automated dispenser) and solid feed (i.e. concentrates, hay, straw, and maize silage in separate troughs) (PREF, n=24), or fed a reference veal diet (i.e. max 500 g DM solid feed mixture and max 22 l/calf/d milk replacer) (REF, n=16). At 3 and 6 months, calves, in groups of 4 (10 groups) visited a test pen for 1 week where individual intake was automatically recorded 24 h/d via a computer. In this pen, calves were instantaneously scanned at 2 min intervals for 30 min every 2.5 h from 7:30 to 18:00, twice. Intake was corrected for metabolic weight. Milk intake was lower at 6 than 3 months ($P=0.024$) and higher in REF than PREF ($P=0.004$). Solid feed intake in PREF reached over 12 times EU minimum requirement (i.e. 250 g). PREF calves ate more concentrates than other feed types ($P<0.01$), but spent more time at the hay trough ($P<0.01$), and from 3 to 6 months, increased concentrates and maize inatkes ($P<0.01$). At 3 months, PREF selected the following proportion of milk, concentrates and roughage: (mean±SD %) 33±18, 36±20 and 31±13. At 6 months, PREF conserved the roughage proportion (27±7), but increased concentrates (56±11) at the expense of milk (17±8). This is in line with calves switching from monogastrics to ruminants. REF calves displayed more AOB and less ruminating than PREF calves ($P<0.01$), but no play behaviour differences were found ($P>0.05$). In PREF, no relationships were found between individual intake of different components and behaviour. Providing a diet choice improved calf behaviour and welfare, regardless of individual differences in intake.

Involving students in animal welfare blogging: the Animalogos experience

Olsson, I. Anna S. [1]*, Galhardo, Leonor*[2]*, Franco, Nuno H.*[1] *and Magalhães-Sant'ana, Manuel* [1,3]*,* [1]*IBMC – Instituto de Biologia Molecular e Celular, Universidade do Porto, Laboratory Animal Science, Rua Campo Alegre 823, 4150-180 Porto, Portugal,* [2]*Instituto Superior de Psicologia Aplicada, Unidade de Investigação em Eco-Etologia, Rua Jardim do Tabaco 34, 1149-041 Lisbon, Portugal,* [3]*Escola Universitária Vasco da Gama, Department of Veterinary Sciences, Quinta de S Jorge, Estrada da Conraria, 3040-714 Coimbra, Portugal; olsson@ibmc.up.pt*

Blogging is one of the many features of the internet-based communication means known as Web 2.0 which enables users to contribute with own resources. When we launched Animalogos (animalogos.blogspot.com) in December 2009, our main aim was to provide a Portuguese-language professional perspective on animal welfare and ethics, issues previously almost exclusively commented on by animal rights and animal protection NGOs. In 2010 we began integrating the blog into teaching. Students in the animal ethics discipline of two courses (a postgraduate course in animal welfare and an undergraduate course in veterinary medicine) develop discussions on the blog and write essays for later publication. For examination purposes, the texts thus produced by the students are evaluated as a short written essay. In 2011/12, we asked formal feedback through an anonymous online questionnaire distributed to students in the postgraduate course (16/21 respondents). While previously not very familiar with blogs (2 read blogs ≥ once a week, 12 read some type of blogs once a week, 4 never read any type of blogs), the students formed an overall positive view of Animalogos as regards accessibility (4.4; maximum score 5), credibility (4.2) and impartiality (3.6). All respondents found this type of examination an overall positive (10 respondents) or very positive (6 respondents) experience. Most respondents expressed that they would like to be examined using this approach in other disciplines in the course (12 yes, 4 don't know/don't answer). In terms of blog dynamics and visibility, the number of page views doubled during the periods when the students used the blog in teaching and the public dialogues in which the students participate correspond to approximately half of the discussion on the blog. We present this as an example of how Web 2.0 can be used to bring animal welfare and ethics teaching into the public domain.
'Biosense – Science engaging society: Life sciences, social sciences and publics' (PTDC/CS-ECS/108011/2008 – FCOMP-01-0124 -FEDER-009237) is a project financed by FEDER Funds through 'COMPETE – Programa Operacional de Fatores de Competitividade' and by national funds through FCT (Fundação para a Ciência e Tecnologia).

Transgenerational effect of maternal deprivation in Japanese quail (*Coturnix c. japonica*)

Le Bot, Océane, Pittet, Florent, De Margerie, Emmanuel, Lumineau, Sophie and Houdelier, Cécilia, University of Rennes, UMR 6552, Rennes, France; le.bot.oceane@gmail.com

In mammals and birds, a maternal deprivation in early life, which is common in intensive livestock, strongly influences behavioural development of young animals, for short and long terms. Maternal deprivation also seems to affect the development of the next generation in mammals. Because such effects are rarely studied in birds, we analyzed the transgenerational effect of maternal deprivation in Japanese quail. First, we determined the behavioural characteristics of females hatched from eggs laid by mothered females (M females) and non-mothered females (NM) (emotional and social characteristics, cognitive abilities in a spatial task, maternal behaviour). Then, we followed the behavioural development of young brooded by those females (4 chicks per female). We used Mann-Whitney non-parametric tests to compare the two sets of females. M (n=22) and NM (n=22) females presented behavioural differences in adulthood, particularly regarding global activity. Indeed, NM females were less active (number of steps in openfield test: M=78.6±17.9, NM=32.0±8.5, P=0.009). Then, NM females showed a higher social motivation than M females (conspecific approach latency: M=122.6±9.5 s, NM=94.5±12.5 s, P=0.022). Moreover, they showed lower cognitive abilities in a detour test (success rate: M=77.3%, NM=22.7%, P=0.006). Maternal behaviour of M and NM females seemed similar, although chicks brooded by NM females maintained greater distance from their mother at the end of the mothering period. Finally, chicks brooded by M and NM mothers were not different after emancipation. We showed here that an early maternal deprivation has an impact on behavioural characteristics of genetic daughters from these non-mothered females, in spite of few impacts on their maternal behaviour. Our study highlighted for the first time a transgenerational effect of such a deprivation in birds.

Intergenerational effects of heat stress on the response of the HPA-axis to acute heat stress in Japanese quails

Henriksen, Rie[1], Rettenbacher, Sophie[2] and Groothuis, Ton[3], [1]Linköping University, AVIAN Behavioural Genomics and Physiology Group, Campus Valla, 581 83 Linköping, Sweden, [2]University of Veterinary Medicine, Veterinaerplatz 1, 1210 Vienna, Austria, [3]University of Groningen, Behavioural biology, Nijenborgh 7, 9747 AG Groningen, the Netherlands; riehe@sol.dk

During the sensitive period of embryonic development changes in the prenatal environment caused by maternal stress shapes how an animal will respond to stress later in life. In most studies so far, it is often unclear how the effects of maternal stress experienced early in life relate to the postnatal test environments in which the offspring's phenotype is assessed. In this experiment we tested whether effects of maternal stress on the offspring HPA-axis could be modulated by the offspring's own postnatal environment. We heat stressed (33-35 °C, 7 hours/day) 18 female Japanese quails for 3 weeks, and kept other 18 female Japanese quails at control temperature (22 °C, 24 hours/day). Eggs were collected from both groups of females and hatched. Half of the offspring from each mother were heat stressed (33-35 °C, 7 hours/day) and the other half were housed at control temperature, which created four offspring groups (1, offspring of CONTROL-mothers raised in CONTROL-rooms; 2, offspring of HEAT-mothers raised in CONTROL-rooms; 3, offspring of CONTROL-mothers raised in HEAT-rooms; 4, offspring of HEAT-mothers raised in HEAT-rooms). At sexual maturity the offspring's cort response to 15 min exposure to 40 °C ambient temperature (heat shock, HS) was measured in all offspring. For statistics a combined hierarchical- and cross-classified linear mixed model was built in MLwiN using Markov chain Monte Carlo. Offspring of HEAT-mothers raised in HEAT-rooms had the highest response to 15 min HS ($F0_{Treatment} \times F1_{Treatment}$; $\chi 2=4.453$; $P=0.035$), whereas offspring of CONTROL-mothers raised in HEAT-rooms as the only group experienced an increase in cort during the recovery period ($F0_{Treatment} \times F1_{Treatment}$; $\chi 2=6.229$; $P=0.013$). These results demonstrate that the response of the HPA-axis is affected by maternal stress but the direction of the effects and whether they are expressed are dependent on the offspring's own environment.

The effect of lighted incubation on growth and pecking behavior in broiler chickens

Dayıoğlu, Miray and Özkan, Sezen, Ege University, Department of Animal Science, 35100 İzmir, Turkey; sezen.ozkan@ege.edu.tr

Although feather pecking may become a problem which is detrimental to birds' welfare, gentle feather pecking is known to be a part of normal behavioral repertoire influencing social exploration of young chicks and could be modified by lighting during incubation. The aim of this study was to investigate the effect of lighting during incubation on post-hatch pecking behavior and growth of broilers. Ross-308 eggs were incubated under 16L:8D program (white fluorescent, 300 lx) and in darkness. Hatched chicks (n=168) were reared in the 8 floor pens (4 pens/treatment; 21 chicks/pen; 12.6 chicks/m^2) under 16L:8D fluorescent lighting (20 lx). Birds were weighed on d 0, 7, 21 and 35. Gentle and severe feather pecking, ground pecking, and pecking own feathers were recorded during 10 min observation periods for each pen on d 5, 7, and 24 within three sessions, morning (9:00-11:00), afternoon (14:00-16:00), and evening (22:00-24:00) hours. Total pecking was the sum of records. Data were presented as frequency per bird (act/bird) averaged across all birds in the pen. Gain and behavior data were subjected to ANOVA. Statistical significance is based on $P \leq 0.05$. Chicks from lighted incubation gained 2.6 g/d more than dark incubated ones between 0-7 d. Gains for 0-21 and 0-35 d didn't differ with treatment. Lighted incubation resulted in higher gentle feather pecking (0.46 and 0.25±0.04) and total pecking (7.54 and 6.74±0.05) in broilers as compared to the dark incubation. Observation time and age affected total pecking and all forms of pecking except ground pecking which didn't differ among sessions. Age by observation time interaction was significant for some forms of pecking behavior. The higher total pecking and gentle feather pecking accompanied by faster early growth in broilers from lighted incubation could be attributed to increased social exploration which may indicate a better adaptation to post-hatch environment.

Foraging development with and without the mother hen and with a tutor chick

Gajdon, Gyula K.[1] *and Stauffacher, Markus*[2], [1]*University of Veterinary Medicine Vienna, Medical University of Vienna, University of Vienna, Messerli Research Institute, Veterinärplatz 1, 1210 Vienna, Austria,* [2]*ETH Zurich, Institute of Agricultural Sciences, Universitätsstrasse 2, 8092 Zurich, Switzerland; gyula.gajdon@vetmeduni.ac.at*

The aim of the work presented here is to investigate the social dynamics of early foraging in domestic chicks (*Gallus gallus*) in groups of varying social structure. For that purpose we gave each of 24 groups of 4 chicks and their hen daily access to a foraging area. We also kept 16 groups of 6 chicks, each, together with a hen, or with a chick 3 days older, or with another chick of same age as its group mates, and offered food bowls the colour of which changed during four days of the first week of life. Food bowl and particle preferences of one focal chick per group were recorded in a test arena on day 9 of life after tutors were removed. Results show that the proportion of the chicks' food cache discloses started to raise only after the first week of life (Friedman test, n=20, P<0.001). Results from the test arena showed that more focal chicks first reacted to particles in bowls of the colour presented during day 3-5 when they were kept together with a hen (75%) or an older chick (79%) than with a group mate of same age (38%; Person χ^2-test, P=0.0016 and P=0.038, resp.). Chicks of both treatments without hen did not differ from each another in particle preferences, but both pecked twice as long at novel particles (gravel) that were similar to familiar food than those kept together with a hen (ANOVA, P=0.018).We conclude that learning is adapted to the natural social dynamics. Imprinting-like learning of food site can be tutored by an older chick, but not the conservative effect of a hen on particle pecking preferences. Thus, our study revealed substantial differences in foraging development despite ample opportunity for individual learning.

Effects of broody hen after hatching on behavior of pullets

Tanaka, Toshio, Shimmura, Tsuyoshi, Fujino, Saori and Uetake, Katsuji, Azabu University, Animal Science and Biotechnology, 1-17-71 Fuchinobe, Chuo-ku, Sagamihara-shi, 252-5201, Japan; tanakat@azabu-u.ac.jp

We previously reported the details of maternal behaviour, the effects of broody hens on the behaviour and character of chicks, and the relationship between maternal behavior and the character of chicks. The objective of the present study was to clarify the lasting effects of brooding by mother hens on the behavior of pullets. Five groups of five chicks each were reared by broody hens and five groups of five chicks were provided with a heater for 28 days. After that, all groups were moved to pens. The behavior of all pullets was recorded using scan sampling at 5 min intervals for 4 h per day at 9 and 16 wks of age. An open field test for evaluating fearfulness was also conducted at both ages, and the number and latency of vocalizations and steps taken were recorded for 10 min. Body weight and feather score were recorded after the test. The data was analyzed using repeated measure ANOVA. The brooded pullets spent more time on non-active behavior ($P<0.05$), but less time on threatening and fighting behaviors ($P<0.1$) than non-brooded chicks. In the open field test, the total duration and numbers of vocalizations and steps taken in both groups were almost the same. On the body weight and feather score, no significant differences between both groups were found. In conclusion, according to our previous report, maternal behavior is closely related to development of thermoregulation, and the behavioral development of chicks was promoted and fearfulness was decreased remarkably by the provision of broody hens, but the lasting effects might not be serious.

Behavioural ontogeny in captive giant panda cubs (*Ailuropoda melanoleuca*) using chronoethology as behavioural monitoring tool

Pertl, Martina[1,2], *Pfistermüller, Regina*[1], *Kratochvil, Helmut*[2] *and Krop-Benesch, Annette*[3], [1]*Tiergarten Schönbrunn GmbH, Maxingstraße 13b, 1130 Vienna, Austria,* [2]*University of Vienna, Department for Evolutionary Biology, UZA 1, Biozentrum, Althanstraße 14, 1090 Vienna, Austria,* [3]*Vectronic Aerospace GmbH, Carl-Scheele-Straße 12, 12489, Germany; martina.pertl@chello.at*

Despite the general interest in the rhythmical organisation of mammal life, only few studies have examined the development of this timekeeping system. This timekeeping system, or biological 'clock', allows the organism to anticipate and prepare for changes in the physical environment. Applied chronoethology is a very comprehensive method of behavioural monitoring and is based on the fact that physical and behavioural responses to illness or social environmental are not only mirrored in behaviour changes itself but also in changes of behavioural pattern and rhythms. By using long-term day and night observations chronoethology can help detect aberrations from behavioural patterns that otherwise may remain unverifiable. The structure of circadian rhythms in giant panda can serve as a case study. Here, we provide the very first results on the behavioural ontogeny of two giant panda cubs, born at the Vienna Zoo in 2007 and 2010, including the development of their activity rhythm and time budgets. Chronoethograms were used to display the behavioural development and to visualize behavioural shifts in activity rhythm and time budgets. The results revealed no significant ($P>0.05$) circadian rhythms in the first weeks after birth and that they do not develop until the third and fourth month. Whereas the activity of the cubs is evenly distributed over 24 h within the first month, they become increasingly diurnal thereafter. Adults, in contrast, show two activity bouts (mainly feeding behaviour) between 8 am - 10 am and 3 pm - 6 pm. We also investigate whether zoo-specific external factors (e.g. animal keepers, zoo visitors, medical examinations etc.) influence the activity rhythm of the cubs by descriptive analyses. Although no effect of the animal keepers on the cub's behavioural patterns can be detect in the chronoethograms, they clearly influence the behaviour of the dam due to their daily routine.

The effects of handling and open field test on vocalization in growing puppies

Vostatková-Koursari, Katerina[1], Baranyiová, Eva[2], Štarha, Pavel[3] and Bednář, Josef[4], [1]on maternity leave, Karousades, Agios Ioannis, 490 81 Corfu, Greece, [2]Institute of Tropics and Subtropics, Czech University of Life Sciences, Kamýcká 129, 165 21 Prague, Czech Republic, [3]Faculty of Mechanical Engineering, Brno University of Technology, Technická 2896/2, 616 69 Brno, Czech Republic, [4]Faculty of Mechanical Engineering, Brno University of Technology, Technická 2896/2, 616 69 Brno, Czech Republic; ebaranyi@seznam.cz

Our aim was to study the effects of neonatal handling on behaviour of puppies exposed to short-term separation from their mother and littermates. Vocalizations were used as indicator of emotionality and studied in puppies exposed to handling and repeated open field (OF) tests. Of 19 beagle puppies (4 litters, 6 males, 13 females), ten were given 5 min of individual handling (separated from the litter) from d 1 to 28 invariably at 08:00. Controls (n=9) were not handled except for being weighed once a week but could see, hear and smell the caretakers. Three 15-min tests were carried out on days 21, 35 and 56 after birth in OF arena 0.6×2×2.4 m and video recorded. From the recordings, vocal behaviours of all puppies were evaluated using our original software. We analyzed latencies to first vocalizations and their total durations using General Linear Models with the above factors versus day, sex, handling, and litter. The latency to vocalization was significantly longer (DF=1, F=6.65, P=0.013) in handled puppies at all ages. The duration of vocalization was significantly ($P<0.05$) shorter in handled puppies only on d 21 (as shown by Mann-Whitney U-test and Student t-test) but GLM indicated a decrease (DF=2, F=4.75, P=0.013) of this variable with advancing age. In conclusion, handling of puppies increased their latency to vocalization (presumably alleviating their anxiety when alone). Moreover, duration of vocalization was affected by age in all puppies, indicating habituation to OFT. Both handling and habituation thus delayed and decreased distress vocalizations of puppies in their early postnatal period. (Supported by VFUB IGA grant 34/2006.)

Early feather pecking as a predictor for feather damage at adult age in laying hens

De Haas, Elske N.[1], Groothuis, Ton G.[2], Kemp, Bas[1] and Rodenburg, Bas T.[3], [1]Wageningen University, Adaptation Physiology Group, De Elst 1, 6708 WD Wageningen, the Netherlands, [2]University of Groningen, Behavioural Biology, Centre for Behaviour and Neuroscience, Nijenborgh 7, 9747 AG Groningen, the Netherlands, [3]Wageningen University, Animal Breeding and Genomics Centre, De Elst 1, 6708 WD Wageningen, the Netherlands; elske.dehaas@wur.nl

Feather pecking (FP) is a behavioural problem in adult laying hens. However, FP may already develop during early life i.e. rearing. We studied FP during rearing on commercial farms and whether FP during rearing affected feather damage during laying. Eight Dekalb White (DW) and eight ISA brown flocks (ISA) were used. We recorded gentle and severe FP (GFP and SFP resp.) in each flock (flock size ±8,000 birds) by 2×20 minutes of observations of 20 hens in a visual space of 1 m^2, at one, five and ten weeks of age. Feather damage was assessed on 20 hens/flock based on the Welfare Quality scoring method (sum of damage to neck, back and belly, total scale 0-2) at five, ten, 15 and 40 weeks of age. Data was averaged per flock and analyzed with a general linear model including hybrid and age, and Pearson correlations were calculated between ages. The number of feather pecks per 20 min (both gentle and severe) was highest at five weeks (GFP 62.88 pecks/20 min, SFP 8.4 pecks/20 min) compared to one (GFP:13.4 pecks/20 min, SFP: 2.8 pecks/20 min) and ten weeks (GFP: 36.0 pecks/20 min, SFP: 2.9 pecks/20 min, $P<0.01$). Feather damage was generally low throughout rearing, but was higher for DW hens at 15 weeks than for ISA hens (0.77 vs. 0.67±0.07, $P<0.05$). Only for the DW hens, severe FP at five and ten weeks was positively correlated with feather damage at ten, 15 and 40 weeks of age (all correlations $r>0.6$, $P<0.05$). For both hybrids, feather damage at ten weeks of age was correlated with feather damage at 40 weeks of age ($r=0.53$, $P<0.05$). These results show that both FP and feather damage during rearing were correlated with feather damage during the laying period and could be used as a predictor for feather pecking during laying in commercial flocks.

Socialising gilts to cope as young sows in grouphousing

Rasmussen, Hanne M., Danish Agriculture and Food Council, Pig Research Centre, Axelborg, Axeltorv 3, 1609 Copenhagen V, Denmark; hmr@lf.dk

The ability of gilts to cope as gestating in group housing may be affected by earlier experience with animals larger than themselves. We investigated the effect of socialising gilts at 6 months of age on behaviour on two commercial farms. Gilts (n=421) were socialised at around 6 months of age by letting 30 gilts from a static group mingle with a group of 120 gestating sows for two weeks. A gate was opened, so the pen of the gilts and the pen of the sows now formed one common pen. Control gilts (n=430) had no former experience with gestating sows as they entered the gestation pens after insemination. After insemination, the gilts were either housed in a dynamic group of gilts or in a dynamic group of sows and gilts. Gilts resting in the dunging area of the pen were registered once per week for the first five weeks after introducing the gilts to the gestation pens. Gilts that did not enter the feeding stations to eat their daily ration were registered for the first five weeks in the gestation pen. All data were analysed using a logistic regression model. The proportion of socialised gilts lying in the dunging area of the pen was lower (0.20 vs. 0.27; $P<0.01$) compared to control when housed in pens with older sows. When the gilts were housed in groups of gilts, no significant difference was found (0.16-0.20; $P=0.16$). Socialising of young gilts exerted enduring effects on the lying behaviour of the gilts when housed with older sows in a commercial herd.

Effect of early social interaction on maternal recognition, welfare and performance of piglets

Figueroa, Jaime, Temple, Deborah, Solà-Oriol, David, Pérez, Jose F. and Manteca, Xavier, Universitat Autònoma de Barcelona, Departament de Ciència Animal i dels Aliments, Facultat de Veterinària, 08193 Bellaterra, Barcelona, Spain; xavier.manteca@uab.es

In natural conditions litters from different sows have contact early in life. However, in intensive productions social interactions are restricted during suckling period. The aim of this experiment was to determine if social interaction between litters before weaning affect maternal recognition, performance and behaviour of piglets before and after weaning. During lactation (28 d), fourteen sows and their litters were randomly allocated into two groups: CG (contact group) and NCG (no-contact group). In CG, 4 litters were housed in pairs from the 2nd-of-birth until weaning, whereas litters of the NCG were kept under commercial conditions. Suckling behaviour was recorded in CG piglets. Social and exploratory behaviour as well as the prevalence of wounds were compared between CH and NCG piglets after weaning. Body weight was measured at birth, at weaning and one week after weaning in order to calculate the average daily gain (ADG). Generalized Linear Model procedure of SAS® was used to analyse the data. During lactation, most CG piglets (94±10%) suckled from their own mother. After weaning, NCG piglets showed more aggressive interactions (17.8 vs. 6.8%; $P<0.001$) and a lower occurrence of positive social interactions (4.6 vs. 11.0%; $P<0.001$) than CG piglets. Prevalence of severely wounded piglets was significantly higher in the NCG than in the CG (14.6 vs. 1.2%; $P=0.03$). Explorative behaviour, ADG during lactation (215 vs. 200 g) and after weaning (95 vs. 114 g) were not affected by treatments ($P>0.5$). However, sows of the CG tended to consume more lactation feed (ADFI; 6,140 vs. 5,621 g; $P=0.07$) but no differences were observed in their body condition or body weight losses. Group housing before weaning may facilitate early interaction between conspecific animals increasing their welfare due to a reduction of social stress after-weaning without affecting performance or normal feeding behaviour during lactation.

Welfare in the wild: personality-dependent effects of noise on breeding birds

Naguib, Marc[1,2], Van Oers, Kees[1], Braakhuis, Annika[1], Griffioen, Maaike[2] and Waas, Joseph R.[3], [1]Netherlands Institute of Ecology (NIOO-KNAW), P.O. Box 50, 6700 AB Wageningen, the Netherlands, [2]Wageningen University, Department of Animal Sciences, Behavioural Ecology Group, the Netherlands, [3]Waikato University, Biological Sciences, Hamilton, New Zealand; marc.naguib@wur.nl

Anthropogenic noise has increased dramatically due to significant expansion of human activities throughout the world. The increase can have substantial implications for animals that communicate acoustically, like birds. Impacts of noise may be affected by spectral characteristics, but this impact might differ between individuals and thus be linked to the personality of the individuals that are exposed to environmental noise. Here, we investigated the influence of noise, differing in spectral characteristics, on parental nest box visits and nestling begging in a field population of personality-typed great tits (Parus major). Exploration behaviour of wild subjects was tested in a standardized novel environment, to obtain an operational measure of personality. When nestlings were 10 days old, we exposed nest boxes of great tits to three treatments: (1) noise that masked the frequency-range of nestling begging calls; (2) noise that did not mask the frequency-range of the calls; and (3) a silent control. Both types of noise significantly reduced parental nest box visits and nestling begging ($P<0.05$, GLMM), indicating that also noise that is not interfering with communication but may affect an animal's behaviour and attention in more general terms, with implications for breeding behaviour. Avoidance of noise by parents depended on their personality, but the effect was in opposite directions for males and females ($P<0.05$). Our results provide insight into how individuals differ in their reaction towards disturbance and highlight how stressors can have different welfare implications, dependent on an individual's behavioural characteristics.

Relationship between behavioural traits and residual feed intake in beef cattle

Haskell, Marie J.[1], Turner, Simon P.[1], Hyslop, Jimmy[1], Bertin, Laurence[2] and Wall, Eileen[1], [1]SAC, West Mains Road, Edinburgh, EH9 3JG, United Kingdom, [2]Bordeaux Sciences Agro, 1 Cours du Général de Gaulle CS 40201, 33175 Gradignan Cedex, France; marie.haskell@sac.ac.uk

Increasing feed efficiency or residual feed intake (RFI), in growing animals is seen as a viable method of reducing the greenhouse gas emissions associated with beef production systems. Residual feed intake (RFI) is calculated by comparing the inputs (feed intake and feed composition) to the outputs (liveweight (LWT) gain and back-fat). However, the behavioural characteristics of the animal, particularly its level of activity and reactivity to challenge, may influence growth efficiency. Ninety Limousin-sired crossbred steers (starting LWT=391; SE=5 kg), housed in four groups in indoor pens were used to investigate this. Steers were weighed once per week and their feed intake measured with automatic feeders over an 8-week test period, with back-fat measures taken at the start and end of this period. In the last 4 weeks, flight speed and crush scores were measured 3 times and pedometers (IceTag™) were attached to the steers' hind legs to assess activity. Number of steps, total durations of standing and lying and number of standing and lying bouts were calculated from this data. Linear Regression models were used to model RFI, and to calculate a score for each steer. REML was used to relate this score to the behavioural variables. The results showed that the mean daily number of steps taken significantly contributed to the prediction of feed intake ($t=2.11$; $df=75$; $P<0.05$). Mean daily steps was also related to RFI ($F=4.13$, $df=1$, $P<0.05$), with animals showing higher numbers of steps/day being less efficient. There was no relationship between crush score, flight speed or duration or frequency of standing or lying bouts with RFI ($P>0.05$). This suggests that the animal's activity is related to growth efficiency and using pedometer-based data can improve RFI predictions. However, any measures taken to improve efficiency may potentially decrease animal activity. The potential impact on animal behaviour and welfare should be seriously considered.

Entrance order and side preference of buffalo cows in the milking parlour

De Rosa, Giuseppe[1], Polikarpus, Annemari[2], Grasso, Fernando[1], Pacelli, Corrado[3] and Napolitano, Fabio[3], [1]University of Naples Federico II, Department of Soil, Plant, Environment and Animal Production Sciences, Via Università 133, 80055 Portici (Napoli), Italy, [2]Estonian University of Life Sciences, Institute of Veterinary Medicine and Animal Sciences, Kreutzwaldi 62, 51014 Tartu, Estonia, [3]University of Basilicata, Department of Animal Science, Via dell'Ateneo Lucano 10, 85100 Potenza, Italy; giderosa@unina.it

Fifty-seven primiparous buffalo cows were used to assess the consistency of entrance order and the preference for a side of the milking parlour. Animals were taken to a waiting area and free to choose their position. They were milked in a 5×5 auto-tandem milking parlour. For each cow, entrance order into milking parlour, side where she was milked and milk yield were recorded. These data were derived from computerised cows' identification. The analysis of data was conducted on 130 morning milkings. The consistency of the entrance order was computed using the Kendall concordance coefficient. The correlation between mean milk yield and mean of entrance order was calculated using the Spearman correlation coefficient. For each cow, side preference in the milking parlour was assessed using the c^2 one sample test, with 50% as expected frequencies. However, it was considered a side preference only if an animal chose a particular side for more than 80% of the milking sessions. For the animals showing side preference, the preference for one stall out of the 5 available in a side was assessed using the c^2 one sample test, with 20% as expected frequencies. It was considered a stall preference only if an animal chose a particular stall for more than 40% of the milking occurred in the preferred side. A constancy of the entrance order into milking parlour was found ($W=0.658$; $P<0.001$). More productive cows tended to enter the milking parlour first ($r_s=-0.221$; $P<0.10$). Twenty-seven cows (47.4%) preferred the right side of the milking parlour, 25 the left side (43.8%) and the remaining 5 (8.8%) showed no preference. Forty-three out of 57 animals were milked in 99-100% occasions in the same side. Nine of the animals showing a side preference also exhibited a stall preference. Our results showed that buffaloes can present marked entrance order consistency and side preference.

Measuring initiative and the assessment of temperament in trainee guide dogs

Harvey, Naomi[1], Craigon, Peter[2], Roberts, Rena[2], Blythe, Simon[2], David, Grice[2], Gary, England[1] and Lucy, Asher[1], [1]The University of Nottingham, School of Veterinary Science & Medicine, Sutton Bonnington, Loughborough, LE12 5RD, United Kingdom, [2]The Guide Dogs for the Blind Association, Hillfields, Burghfield Common, Reading, Berkshire, RG7 3YG, United Kingdom; svxnh@nottingham.ac.uk

Working guide dogs need to display many behavioural qualities including the ability to navigate new obstacles or find detours from usual routes. This studies aims were: (1) to trial a method of measuring initiative; (2) to trial a questionnaire devised to assess temperament. We defined initiative as 'an unaided and active attempt to resolve a novel problem'. To test initiative a controlled detour based maze test was developed, 45 trainee dogs were trained to use one route of a maze, that route was then blocked, an alternative opened and responses recorded. To trial potential measures of temperament a questionnaire was constructed, combining elements of the C-BARQ and dog-ADHD questionnaires and completed by the dogs trainers. Exploratory PCA was applied to the questionnaire data and nine stable, interpretable components emerged: Distractibility; Stranger fear/aggression; Non-social fear/anxiety; Trainability; Animal chase; Immaturity; Excitability; Vehicle Chase and Body Sensitivity. An 'Initiative' component composed of specific items from the other traits was also identified via correlational analyses. All components had high internal reliability (α 0.7 to 0.9). Time taken to completion of the detour task on the first trial and direction chosen in the second trial were considered as possible candidate measures of initiative. Significant associations (ANOVA $P<0.05$) with success or withdrawal were found with seven of the questionnaire components, with withdrawn dogs scoring significantly higher on; Distractibility; Stranger fear/aggression; Non-social fear/anxiety; Animal chase; Vehicle Chase, and lower on: Trainability and Initiative. Dogs withdrawn for health problems (skin & eye related) were included in the analysis and found to score significantly higher ($P<0.01$) than qualified dogs for the component Non-social Fear/Anxiety. In summary, the questionnaire appeared to be a valuable tool for identifying behaviour important to suitability to the guiding role and may also indicate a possible link between certain health issues and anxiety in dogs.

Personality and well-being in Scottish wildcats (*Felis silvestris grampia*)

Gartner, Marieke C., University of Edinburgh, Psychology, 7 George Square, EH8 9JZ, United Kingdom; m.c.gartner@sms.ed.ac.uk

Personality is one of the strongest, most consistent predictors of well-being in humans. In humans and some nonhuman primates, well-being is positively correlated with Extraversion and Agreeableness, and negatively with Neuroticism. In addition to genetic properties, well-being is also attributable to environmental effects. Studying well-being, then, may have important implications for captive management. There has been little work done on felids and personality and well-being. Research suggests that personality may have implications for increasing welfare in zoos, from altering environments to decreasing stress. This study looked at the Scottish wildcat, a critically endangered species. Eight caretakers completed personality and well-being surveys on 23 wildcats across three zoos. Each zoo had similar outdoor enclosures, meant to mimic natural habitat. The personality survey (42 traits rated on a Likert scale) was based on previous felid and primate research. The well-being survey was identical to one used on primates. Four questions on mood, sociability, achieving goals, and how happy a rater would be to be the animal in question were rated on a Likert scale. The mean reliability of individual personality ratings was 0.35 (ICC 3,1), and of mean ratings was 0.55 (ICC 3,k). The mean reliability of individual well-being ratings was 0.28, and of mean ratings was 0.53. Principal-components analysis revealed three personality factors: Dominance, Agreeableness, and Neuroticism. Neuroticism was negatively correlated with well-being ($r=-0.677$, $P=0.001$). It was related to the mood ($r=-0.616$, $P=0.003$) and sociability ($r=-0.811$, $P=0.000$) measures of well-being. Well-being was unaffected by age or sex. To address the issues associated with wildcat Neuroticism, specifically less sociability and increased negative affect, it is likely that housing this type of wildcat alone would be beneficial to its well-being. These animals should also be given more places to hide, in order to give them some control over their environment.

Are breed and sex associated with the likelihood of stabled UK leisure horses displaying stereotypic behaviour?

Hockenhull, Jo[1] and Creighton, Emma[2], [1]University of Bristol, School of Veterinary Sciences, Langford House, Langford, Bristol BS40 5DU, United Kingdom, [2]Newcastle University, School of Agriculture, Food and Rural Development, Claremont Road, Newcastle upon Tyne NE1 7RU, United Kingdom; Jo.Hockenhull@bristol.ac.uk

Leisure horses form the majority of the UK equine population. Over the last decade there has been a change in the way these horses are managed with a growing trend towards year-round stabling. The length of time horses spend stabled has been associated with the expression of stereotypic behaviours. Previous studies focusing on performance horses have found a relationship between the breed of the horse and likelihood of it displaying oral and locomotor stereotypies. This study used data generated as part of a larger project to explore whether the breed and sex of stabled UK leisure horses are associated with the likelihood of displaying stereotypic behaviours. Horse-level data were collected via Internet surveys from a convenience sample of UK leisure horse owners. Logistic regression analyses were used to investigate whether breed and sex were associated with the performance of locomotor, oral and pre-feeding locomotor (weaving, box-walking, head-nodding) stereotypic behaviour. In comparison to native breeds, thoroughbreds and thoroughbred crosses were associated with an increased likelihood of displaying locomotor (n=371; P=0.025; OR 3.28 (1.16-9.24)) and pre-feeding locomotor stereotypies (n=404; P<0.001; OR 6.40 (2.48-16.51)). No association was found between breed and the performance of oral stereotypies. Mares were associated with an increased likelihood of displaying locomotor stereotypies (n=371; P=0.002; OR 2.59 (1.44-4.67)) compared to geldings or stallions. While causal relationships between breed, sex and stereoptypies cannot be determined from this study, associations were found indicating that thoroughbreds, thoroughbred crosses and mares may be more likely to display these behaviours. These horses maybe more reactive, and thus maybe more likely to employ active coping strategies. However, the lack of association between non-thoroughbred, male horses and stereotypic behaviour should not be taken as an indicator that the welfare of these horses is optimal, rather that they maybe less likely to respond actively to environmental stressors.

Testing an adaptation of the qualitative behaviour assessment (QBA) method to assess cattle temperament

Sant'anna, Aline C.[1,2], Paranhos Da Costa, Mateus J.R.[2], Rueda, Paola M.[2,3], Soares, Désirée R.[2,3] and Wemelsfelder, Francoise[4], [1]Programa de Pós-Graduação em Genética e Melhoramento Animal, Av. José A.A. Martins, 290, Jaboticabal, 14883-298, Brazil, [2]UNESP – Universidade Estadual Paulista 'Julio de Mesquita Filho', Zootecnia, Via de Acesso Prof. Paulo D. Castellane, Jaboticabal, 14884-900, Brazil, [3]Programa de Pós-graduação em Zootecnia, Via de Acesso Prof. Paulo D. Castellane, Jaboticabal, 14884-900, Brazil, [4]Scottish Agricultural College, Edinburgh, EH9 3JG, United Kingdom; ac_santanna@yahoo.com.br

This study aimed to test the use of qualitative behaviour assessment (QBA) method to assess cattle temperament. A total of 335 Nellore young bulls were evaluated once during the handling for weight determination, assessing the QBA and flight speed (FS). The cattle QBA list developed for the EU Welfare Quality project used 20 terms as indicators of animal welfare. The method was adapted, reducing to 12 terms to increase its feasibility. Terms expressing positive and negative aspects of temperament were maintained evenly, as follow: active, relaxed, fearful, agitated, calm, attentive, positively occupied, curious, irritable, apathetic, happy and stressed. For each term there was a visual scale of 125 mm on which one observer scored his qualitative assessment of the animal's expression. FS is an objective method often used to assess cattle temperament, recording the speed (m/s) that an animal exits a crush. Principal Component Analysis (PCA) was used for data analysis, based on the index obtained for each individual in the first principal component (PC1) a new variable was defined, the temperament index (TI). Pearson's coefficient of correlation was calculated between TI and FS. PC1 and PC2 accounted for 51.56% and 10.28% of the total variance, respectively. PC1 had major positive contributions of the terms agitated (0.94) and active (0.88) and major negative contributions of the terms relaxed (-0.94) and calm (-0.89), indicating that as higher TI value, worse is the temperament. FS and TI were positively correlated ($r=0.56$, $P<0.01$), indicating that individuals with high TI are likely to present also high FS. We conclude that the QBA adaptation has potential to assess cattle temperament, since was possible to identify individual variation on this trait. Financial support: FAPESP (09/53608-0).

Qualitative behaviour assessment in sheep: consistency across time and association with health indicators

Phythian, Clare[1], Wemelsfelder, Francoise[2], Michalopoulou, Eleni[1] and Duncan, Jennifer[1], [1]University of Liverpool, Epidemiology and Population Health, Leahurst, CH64 7TE, United Kingdom, [2]Scottish Agricultural College, Sustainable Livestock Systems, Penicuik, EH26 0PH, United Kingdom; sheep-vet@bristol.ac.uk

Qualitative Behaviour Assessment (QBA) is a 'whole-animal' methodology that addresses welfare by scoring an animal's body language using terms such as content, anxious or relaxed. This method has shown to be a reliable and feasible indicator of the on-farm welfare in pigs, cattle, poultry and sheep, however the consistency of QBA over time has not yet been addressed. This study aimed to investigate whether and how on-farm QBA of sheep varies over different seasons of the year and the association with other sheep health indicators. A trained assessor visited 12 farms six times per year at 2 month intervals. At each farm either the entire flock was assessed, or if flocks were large a group of ±100 sheep were selected ad hoc (assuming flock homogeneity). QBA was based on a list of 12 descriptors previously developed for sheep in collaboration with Quality Meat Scotland and the Scottish Society for the Prevention of Cruelty to Animals. Following QBA, the number of sheep with signs of lameness and breech dirtiness was counted. QBA scores from the 6 assessment dates were analysed together using Principal Component Analysis (PCA – correlation matrix, no rotation). The effect of assessment-date on PCA farm scores was analysed with repeated-measures analysis of variance. The association of PCA farm scores with the proportion of lame and 'dirty rear' sheep was examined by Spearman's rank correlation coefficient. PCA distinguished two dimensions: PC1 (48.2% variation) ranging from 'content/relaxed/thriving' to 'distressed/dull/dejected' (summarized as 'mood'), and PC2 (19.81%) which ranged from 'anxious/agitated/responsive' to 'relaxed' (summarized as 'responsiveness'). No effect was found of assessment-date on PC1 ($P<0.31$), indicating the sheep's general mood appeared relatively stable throughout the year. There was an effect of assessment-date on PC2 ($P<0.001$), indicating a possible 'relaxing' effect of young lambs on the sheep's responsiveness. The proportion of lame sheep was negatively correlated with farm scores on PC1 (n=72; rho=-0.59, $P<0.001$), indicating an association of negative 'mood' with lameness. No other significant correlations were found. By demonstrating the relative stability of such sheep farm scores over time, and their meaningful association with lameness, these results further support the reliability and validity of QBA as an indicator used in on-farm assessments of sheep welfare.

On-farm assessment of social behaviour of dairy goats: duration of observation and time of day

Schmied-Wagner, Claudia, Nordmann, Eva, Mersmann, Dorit and Waiblinger, Susanne, Institute of Animal Husbandry and Animal Welfare, University of Veterinary Medicine Vienna, Veterinärplatz 1, 1210 Vienna, Austria; claudia.schmied@vetmeduni.ac.at

On-farm welfare assessment of social behaviour requires a short observation time. We examined, whether 1-h or 2-h observation periods provide comparable information on the frequency of social interactions of dairy goats as does a 6 h observation period, and the potential influence of time of day. On 45 farms with 155±96 dairy goats social behaviour was recorded directly in three blocks of two hours: day1 before evening milking (Afternoon); day1 after evening milking (Evening); day2 after morning milking (Morning). We calculated number of different social interactions for the total of 6 h (Total) and for each block, as well as for the first (Evening1) and second hour (Evening2) of Evening. Spearman rank correlations were calculated. Frequency of agonistic interactions during Total correlated highly with frequency during Evening (with body contact AgoBC: r_s=0.90, without body contact AgoNo: r_s=0.82, both P<0.001), somewhat lower with frequency during Morning (AgoBC: r_s=0.83, AgoNo: r_s=0.79, both P<0.001), but much lower with Afternoon (AgoBC r_s=0.49, P=0.001, AgoNo r_s=0.41, P=0.005). For socio-positive behaviour time of day seems not important as correlations of Total were similar with Afternoon, Evening and Morning (SocPos: r_s=0.88, 0.83, 0.88, all P<0.001). Reducing the observation time to one hour influenced Ago and SocPos differently: Agonistic behaviour correlated stronger with Total in the first hour after milking (Evening1: AgoBC r_s=0.76 P=0.000, AgoNo r_s=0.79 P=0.000), which is mostly equivalent to feeding time. In contrast, for SocPos Evening2 correlated stronger with Total (r_s=0.82 P=0.000) than Evening1 (r_s=0.39 P=0.008). Calculating linear regression models to identify factors influencing AgoBC revealed comparable results for Total and Evening (no change in R^2adj.), but reduction in R^2adj. when using Morning, Evening1 or Evening2 as dependent variable. Our data suggest that reduction of observation time to two hours after evening milking is possible with sufficient reliability for on-farm assessment of dairy goats' social behaviour.

Evaluation of on-farm parameters of behaviours around resting in dairy cattle

Hauschild, Birte, Tremetsberger, Lukas and Winckler, Christoph, University of Natural Resources and Life Sciences, Department of Sustainable Agricultural Systems, Division of Livestock Sciences, Gregor-Mendel-Str. 33, 1180 Vienna, Austria; christoph.winckler@boku.ac.at

Since on-farm recording of lying behaviour is less feasible, measures of resting behaviour in animal-based welfare assessment protocols mostly focus on movements or lying positions. E.g. the Welfare Quality® assessment protocol for dairy cattle includes the time needed to lie down, collisions with the equipment and animals lying outside the lying area. It was the aim of this study to investigate more detailed measures of spontaneous lying down movements including intention movements and a rising score which had previously been developed for tied cows. 17 cubicle-housed dairy herds (mean size 33 cows) in Upper Austria were included. Lying down behaviour was continuously recorded in segments of the barn during 2 to 2.5 hours after the morning milking and rising behaviour was scored for on average 20 animals after the behaviour observations. All observations were carried out by the same observer and data were analysed at herd level using general linear models, Spearman correlations and Chi^2-tests. Farms differed significantly in the time the cows stood with four feet in the cubicles before lying down (range 8.5±12.7-87.1±17.9s, P=0.002) as well as in the duration of the carpal stance phase (P=0.008) but there were no significant differences in the time standing with two feet in the cubicles and the time spent on intention behaviours such as head swinging or sniffing. Correlations between the different phases were low (rs=0.09-0. 29). Rising scores were affected by lameness (3-category-scale; P=0.005) and lactation number (P=0.007) with lame and older animals ($\geq 5^{th}$ lactation) being more likely to show difficulties. Cows that could not be motivated to stand up, however, did not differ in lameness, age or lactational stage. In conclusion, the measures investigated may add information to and be easily incorporated in existing welfare assessment protocols for dairy cattle. Further studies e.g. on repeatability are recommended.

Preliminary results on the application of the Welfare Quality® protocol to assess beef cattle kept under intensive conditions in Spain

Mainau, Eva[1], Magnani, Diego[2], Pedernera, Cecilia[1], Dalmau, Antoni[1] and Velarde, Antonio[1], [1]IRTA, Institut de Recerca i Tecnologia Agroalimentàries, Finca Camps i Armet, s/n, 17121, Monells (Girona), Spain, [2]Università degli Studi di Bologna, Facoltà di Agraria, Dipartimento di Protezione e Valorizzazione Agroalimentare, via Fratelli Rosselli 107, 42100 Reggio Emilia, Italy; eva.mainau@irta.cat

The present study investigates the feasibility of the Welfare Quality® protocol for beef cattle in intensive commercial conditions. Descriptive results from fourteen Spanish intensive farms are presented based on the assessments carried out between March to June 2011. Farms had a mean size of 310 animals (from 81 to 1,274). A total of 1,495 animals were assessed for behavioural measures (agonistic and cohesive behaviours and time needed to lie down), 774 for human-animal relationship (avoidance distance to humans), 2,137 for qualitative behaviour assessment (QBA; based on a set of 20 descriptors) and 912 for clinical measures. The space allowance (mean ± SE) was 54.0±2.18 kg/m^2. Each pen had 1 drinker for 16.7±1.24 animals with a length of 2.9±0.39 cm per animal. The 19.8±8.03% of the drinkers were scored as very dirty and 67.2±7.74% as partially dirty. Each animal showed 0.19±0.007 and 0.14±0.008 hourly rate of agonistic and cohesive behaviours respectively. The time to lie down was 5.5±0.32 seconds. A 28.0±3.24% of the animals avoided human presence further than 1 m, and a 9.2±2.16% could be touched by the assessor. The highest scorings for QBA were obtained in active, calm and inquisitive descriptors and the lowest for frustrated, boisterous and distressed. Moderate and severe integument alterations and lameness were found in 12.9±2.38, 2.9±0.82 and 0.3±0.17% of the animals, respectively. Nasal and ocular discharge was found in a 12.1±1.75 and 6.2±0.81% of the animals, respectively and diarrhoea in a 1.1±0.54%. A 4.54±1.87% of the animals were classified as dirty. The WQ protocol seems feasible in cattle beef and may allow detecting differences between farms.

Application of the Welfare Quality® protocol to small-scale finishing pig farms

Braghieri, Ada[1], De Rosa, Giuseppe[2], Serrapica, Maria[1] and Napolitano, Fabio[1], [1]Università della Basilicata, Dipartimento di Scienze delle Produzioni animali, Via dell'Ateneo Lucano 10, 85100 Potenza, Italy, [2]Università di Napoli Federico II, Dipartimento Scienze del Suolo, della Pianta, dell'Ambiente e delle Produzioni animali, Via Università 133, 80055 Portici (NA), Italy; fabio.napolitano@unibas.it

In order to assess whether household farms may provide higher welfare standards than industrial enterprises, the Welfare Quality® scheme was used. It is based on 26 measures summarised in 12 criteria, 4 principles and a final score ranging from 0 to 100. Five household and five industrial enterprises were monitored. The former were characterised by indoor concrete floor, access to pasture, and a lower number of heads (mean±SD=27.75±22.15); the latter had indoor slatted floor, no access to pasture, and a higher number of heads (110.50±44.89). Comparisons between the two housing systems concerning welfare measures, criteria and principles were performed using the Wilcoxon rank-sum test. The distributions of final scores over the four Welfare Quality® categories ('excellent', 'enhanced', 'acceptable', 'not classified') were assessed using the c^2 test. Household farms showed values higher than industrial farms for the following criteria: ease of movement (Z=2.19, P<0.05), positive emotional state (Z=2.06, P<0.05), expression of other behaviours (Z=2.82, P<0.01) and absence of pain induced by management procedures (Z=3.48, P<0.001). These results can be attributed to the lack of pasture, the lower percentage exploration (Z=2.82, P<0.01) and the higher percentage of tail docked pigs (Z=-3.51, P<0.001) observed in industrial farms. Higher percentages of dirty animals (Z=-2.53, P<0.01), scouring (Z=-2.20, P<0.05) and mortality (Z=-2.04, P<0.05) were also observed in industrial farms. At principle level, household farms showed higher scores than industrial farms for 'good housing' (Z=2.19, P<0.05), 'good health' (Z=2.19, P<0.01) and 'appropriate behaviour' (Z=1.93, P<0.05). As to final score, no farms were rated either 'excellent' or 'not classified'; albeit not significantly, household farms showed a higher percentage of scores 'enhanced' (75%) as compared with industrial farms (50%). Therefore, the Welfare Quality® scheme was able to detect differences between farms despite of the low sample size; in addition, at least at principle level, household enterprises showed higher welfare scores than industrial farms.

Performance of the Welfare Quality® protocol at slaughterhouse in the traditional Majorcan Black Pig production system

Llonch, Pol[1], Fàbrega, Emma[1], Brioullet, Armelle[1], Dalmau, Antoni[1], Jaume, Jaume[2] and Gonzàlez, Joel[1], [1]IRTA, Finca Camps i Armet S/N, 17121 Monells (Girona), Spain, [2]IBABSA, Esperanto 8, 07198 Son Ferriol (Palma), Spain; pol.llonch@irta.cat

The aim of the study was to assess the feasibility of the Welfare Quality® protocol applied to a low capacity slaughterhouse (6,000 pigs slaughtered per year) used for a local pig breed (Majorcan Black Pig; MBP) and to determine the system's strengths and weaknesses in terms of animal welfare. The protocol was performed in two visits at the same slaughterhouse, with a total of 60 pigs assessed. The incidence of dead, sick and panting animals was scored in 4 lorries (15 animals per lorry) at the arrival. During unloading, the percentage of pigs that slipped, fell, showed reluctance to move or turned back and were lame was assessed. In the lairage pens, the stocking density and the percentage of pigs panting, shivering, and huddling were assessed. During conduction to stunning, high-pitched vocalisations were recorded. The stunning effectiveness and presence of lesions on the body, pneumonia, pleurisy, pericarditis and white spots on liver were also assessed. During unloading, 26.7% of MBP slipped and 6.7% fell. Rhythmic breathing and righting reflex, both used as a sign of poor stunning, were observed in 90.0% and 24.2% of the animals, respectively. In 21% of the cases, vocalisations were present. Signs of pneumonia and white spots on the liver were seen in 11.6% and 3.0% of the MBP, respectively. A 2% of the carcasses showed a high score of skin lesions and in 34.7% the presence of lesions was negligible or very low. To apply the protocol some changes were required due to the low number of animals slaughtered per day. However, it can be solved by increasing the number of visits. The poor stunning, suggests that the electrical parameters should be adjusted according to breed particularities. In contrast, the skin lesions had better results than the values obtained in previous studies in big slaughterhouses using the same protocol.

Measuring welfare of finishing pigs from loading at the producer until sticking at a commercial abattoir

Brandt, Pia[1,2]*, Rousing, Tine*[2]*, Herskin, Mette S.*[2]*, Edwards, David*[2] *and Aaslyng, Margit D.*[1]*,* [1]*Danish Meat Research Institute, Danish Technological Institute, Maglegaardsvej 2, 4000 Roskilde, Denmark,* [2]*Department of Animal Science, Aarhus University, Blichers Allé 20, 8830 Tjele, Denmark; pbt@teknologisk.dk*

The market and the consumers show increasing interest in and demands towards animal welfare, and EU regulation (1099/2009) applying from 2013 determines that larger slaughterhouses (>1000 animal units/year) should document animal welfare. Compliance with these requirements necessitates the development of tools for continuous monitoring of animal welfare. The aim of this study was to evaluate different indicators of animal welfare to identify possible measures for future continuous documentation of animal welfare on the day of slaughter. Behavioral, clinical and physiological measures based on the Welfare Quality® Protocol (2009) were recorded. 80 pigs from 4 producers were included and recordings were carried out at loading, unloading, during lairage, prior to stunning, at sticking and post mortem. Behavioral measures recorded as frequencies included 'reluctance to move', 'vocalization', 'falls' and 'aggressiveness'. During lairage 'time to lie down' was recorded by instantaneous sampling. At loading and during lairage, the behavioral recordings were performed by direct observation, whereas at unloading and prior to stunning recordings were performed using video surveillance. Further, lameness, skin damage, heart rate and body surface temperature was recorded, and blood samples were collected at sticking. Post mortem, pH and temperature was measured. A regression analysis was carried out to investigate the effect of a combined index of the behavior before slaughter on the post mortem temperature, pH and blood parameters. The behavior during lairage tended to affect the temperature 45 minutes after sticking (P=0.06). Further, there was an effect of behavior prior to stunning on the concentration of glucose (P=0.02) and a tendency to an effect of the concentration of lactate on behavior prior to stunning (P=0.10) and behavior during lairage (P=0.08). This indicates that temperature and the blood concentration of lactate and glucose might be relevant indicators for documentation of animal welfare on the day of slaughter.

The development of a welfare assessment protocol for captive wolves (*Canis lupus*)

Brand, Lusanne, Kuijpers, Maaike and Koene, Paul, Wageningen University and Research Center, Wageningen Livestock Research, De Elst 1, 6700 AH Wageningen, the Netherlands; lusanne.brand@wur.nl

Zoos are sometimes criticized by society for keeping animals in captivity. It is thus of major importance that zoos guarantee welfare of their animals. Present zoo management provides for increasing and maintaining the welfare potential of zoo animals but often fails when it comes to actual individual welfare management. Therefore, there is a need for an animal-based welfare assessment method. The goal of this study was to develop such a welfare assessment method. Starting with one species as a model: the wolf (*Canis lupus*), valid indicators for wolf welfare were selected showing that social and stereotypic behaviour, housing and food were crucial. In this study also a quick assessment (Tool = behaviour measurements – scan and focal sampling – within 30 minutes) is compared with in-depth study (Detail = behaviour measurements – scan, focal and behaviour sampling – in 2 hours) on individual wolves (n=28) in 5 Dutch zoos. The reliability of the Tool was compared with Detail with Spearman ranks correlations (Rs) and Bland-Altman 'limits of agreement' (LOA) between measurements. This resulted in a Tool design containing exploring, lying together, play social and friendly social interactions as positive animal-based welfare indicators and nervous behaviour, stereotypic behaviour, increased aggressive behaviour, being alone, whimper vocalizations, very low body posture, apathy and self-mutilation as negative animal-based welfare indicators (all Rs, $P<0.05$ and all LOA lower than 15%). In a second study the developed animal-based welfare indicators (the Tool) is developed into a Welfare Quality®-like protocol (by combing appropriate behaviour with good housing, feeding and health) using detailed zoo records and doing detailed observations on the wolves of Ouwehand zoo (Rhenen, the Netherlands). The protocol is applied on 2 other wolf packs in the Netherlands. The conclusion is that the development of a welfare assessment tool for zoo animals is feasible with the wolf as a model.

Can animal-based measures be used in official animal welfare/protection control on Swedish dairy farms?

Holmberg, Mia[1], Winblad Von Walter, Louise[2], Stéen, Margareta[3] and Dahlborn, Kristina[1], [1]Swedish University of Agricultural Sciences, Anatomy, Physiology and Biochemistry, Box 7011, 750 07 Uppsala, Sweden, [2]Swedish Dairy Association, Box 210, 101 24 Stockholm, Sweden, [3]Swedish Centre for Animal Welfare, Box 7068, 750 07 Uppsala, Sweden; mia.holmberg@slu.se

In this project, animal welfare/protection was evaluated at dairy farms in four different regions of Sweden by three different assessment systems: (1) The official animal welfare control in Sweden (OC); (2) The Swedish Dairy Association program 'Ask the cow' (AC); (3) The Welfare Quality® (WQ) assessment protocol for dairy cows. The OC system mainly uses resource-based measures while the AC and WQ systems use animal-based measures. Collection of data was performed by trained assessors, one from each system, on 41 farms. The mean herd size was 65 cows (range 12-268). The project aimed to investigate if the systems ranked the farms in the same order, and which indicators were most important in each system. The Factor procedure (Varimax, SAS 9.2) was used to identify which indicators contributed most to the variation within systems and the CORR procedure was used to identify correlations. The indicators used in AC were explained (60% of variation) by 4 factors and in WQ (62% of variation) by 6 factors. Factor 1 in WQ (positive behaviors) had a negative correlation ($P<0.05$) with factor 1 in AC (dirty cows, lameness) and factor 2 in AC (lesions, lameness). WQ factor 1 had a negative correlation whereas AC factor 2 had a positive correlation with number of remarks in OC. In both WQ and AC high proportions (mean 18% in AC and 50% in WQ) of dirty cows were detected, although only two farms had remarks on dirty cows in the OC. In total (all farms included) there were only 55 remarks on all measured indicators in the OC. The systems rank the farms in different order, partly because the measurements are different. Based on our results animal-based measures are recommended for official control since they enable quantifying for example body condition, cleanliness, lesions and important behaviors.

Different approaches to use animal welfare outcomes for advisory purposes

Temple, Déborah[1], Dalmau, Antoni[2], Velarde, Antonio[2] and Manteca, Xavier[1], [1]Universitat Autonoma de Barcelona, Facultat de Veterinaria. Campus de la UAB., 08193. Cerdanyola del Valles, Spain, [2]IRTA, Finca Camps i Armet s/n, 17121. Monells. Girona, Spain; deborah.temple@uab.cat

Welfare assessment systems can be used for certification purposes and as advisory tools, and their results need to be simple and transparent enough to be clearly understood by farmers. Different approaches were applied for advisory purposes to the results from the Welfare Quality® protocol for growing pigs. First, a benchmarking system was developed by means of Generalized Linear Mixed Models (GLMM) for non-parametric data with low prevalences and count data. The model was applied on 30 intensive farms and upper confidence limits (CL) were established for each animal-based measure (e.g. social negative behaviour: Upper CL=7%; tail biting: Upper CL=2.3%). The methodology allowed the identification of farms with welfare problems in a given population. Second, mean prevalences and distributions of welfare outcomes were obtained on a wide range of commercial farms (91 farms; 5 production systems). GLMM were used to model the data in order to evaluate the effect of the production system and to identify possible causal factors. Important differences between production systems were established based on the majority of animal-based parameters. The odds ratios (OR) reported for several parameters of housing were high when comparing production systems (e.g. OR=67 for pig dirtiness between pigs on straw and those in extensive conditions), suggesting that the production system by itself has an important effect on such welfare problems. In contrast, most health measures did not differ markedly between the five production systems studied as OR included the unity or were close to 1. For many animal-based measures, several causal factors such as feeding system, space allowance, type of floor and growing stage were identified and their respective impact (OR) varied according to the production system.

Scoring egg certification programs in the US: using a science-based computer model to assess hen welfare

Liszewski, Melissa A.[1], De Mol, Rudi M.[2] and Jones, Dena M.[1], [1]Animal Welfare Institute, 900 Pennsylvania Ave., SE, Washington, DC 20003, USA, [2]Wageningen University and Research Centre, Wageningen Livestock Research, P.O. Box 65, 8200 AB Lelystad, the Netherlands; melissa@awionline.org

With the wide variety of animal welfare certification programs for eggs in the United States – each with their own specific standards – it is easy to understand why many American consumers are confused about the way laying hens are raised. Consumers wanting to support more humane farming systems are faced with paying premium prices for welfare certified products from hens not necessarily raised significantly different from the conventional industry. To potentially increase consumer confidence in third-party certification for eggs in the United States and help make these labels more meaningful, we utilized the science-based computer model known as FOWEL to assign welfare scores to different sets of standards within various certification programs (n=6). The FOWEL model calculates scores based on 25 attributes with two or more levels each – such as space per hen, foraging, beak trimming, etc. Weighting factors based on scientific knowledge are combined with attribute levels to calculate a final welfare score for each system. Scores were calculated based on the minimum conditions allowed under each system's standards. Though some producers within each program may independently score higher, the purpose of the analysis was to score the strength of the program itself rather than its individual producers. Results placed the welfare score of United Egg Producers (UEP) Certified standards the lowest with a relative score of 0.00 and the Animal Welfare Approved (AWA) program highest with a relative score of 10.0; absolute scores were 44.33 and 186.37, respectively, with a theoretical maximum of 246. Welfare scores of programs using cage systems were low (UEP Certified and American Humane Certified: Enriched Colony Housing); those of barn and aviary systems without required outdoor access were medium (American Humane Certified: Aviary and Certified Humane); and those with free range/pasture systems were high (USDA Organic and Animal Welfare Approved).

A study of the effect of flooring during pregnancy and lameness on the postural behaviour of gilts in farrowing crates

Calderón Díaz, Julia Adriana[1,2], Fahey, Alan[1] and Boyle, Laura[2], [1]University College Dublin, Dept. of Animal Science, Belfield, Dublin 4, Ireland, [2]Teagasc, Pig Production Dept., Moorepark Research Centre, Fermoy, Co. Cork, Ireland; julia.calderon-diaz@teagasc.ie

This study evaluated the effect of flooring during gestation on postural behaviour of gilts in the farrowing crate and its relationship with lameness. 27 pregnant gilts were loose housed in pens with free access stalls and a slatted loose area in groups of 4. The concrete slats (slat=14.5 cm, gap=2 cm) in the loose area were either uncovered (CON; n=15) or covered by rubber slat mats (RUB; n=12). Gilts were transferred to farrowing crates on day 110 of pregnancy (-5d). Lameness was scored (0 to 5) on -5d and at weaning (28d). Gilts were classified as non-lame (≤1) or lame (≥2). Gilts were videotaped on -5d and before weaning (27d) for 24 h. Videos were sampled instantaneously every 10 min and 4 postures (Standing (St), Sitting (Si), Ventral (VL) and Lateral Lying (LL)) were recorded. Number of postural changes (PC) was also calculated. PC were analysed using PROC MIXED of SAS including flooring, period and lameness. Postures were analyzed using PROC NPAR1WAY. Fifteen (CON=8; RUB=7) gilts were lame on -5d and 11 gilts (CON=5; RUB=6) were lame on 28d. PC were not affected by flooring or lameness ($P>0.05$). Gilts made more PC at 28d compared to -5d ($P=0.001$). Gilts spent >90% of their time lying-down. On -5d RUB gilts spent more time Si (1.19% vs. 0.64%, $P=0.01$) and less time LL (37.73% vs. 49.81%, $P<0.05$) than CON gilts. There was no effect of flooring on time spent in each posture on 27d ($P>0.05$). Lame gilts spent more time St ($P=0.001$) and LL ($P<0.05$) and less time VL ($P=0.01$) compared to non-lame gilts. Behaviour of RUB gilts on entry to the farrowing crate could reflect their efforts to adapt to lying on a more uncomfortable surface. Lameness had a greater influence on the postural behaviour of gilts in farrowing crates than the flooring they were kept on during pregnancy.

The effect of housing during lactation on sow responsiveness

Melišová, Michala[1,2], Illmann, Gudrun[2], Chaloupková, Helena[1,2] and Bozděchová, Barbora[1,2], [1]Czech University of Life Sciences, The Faculty of Agrobiology, Food and Natural Resources, Kamýcká 129, 160 00 Prague Suchdol, Czech Republic, [2]Institute of Animal Science, Department of Ethology, Přátelství 815, 104 00 Prague – Uhříněves, Czech Republic; melisovam@centrum.cz

Sows and piglets are housed during lactation in crates which seriously impair their welfare. The main objections against loose-housing systems are the fear of high piglet mortality and high responsiveness towards stock-person. The aim of the study was to investigate the effect of pens (4.6 m^2, n=20) versus crates (1.6 m^2, n=18) on: (1) sow responsiveness towards playback test of own piglet calls simulating piglet crushing (Crush Test) and towards real crushing; (2) stock-person while handling vocalizing piglet (Handling Test); and (3) on piglet mortality at Day 28 post partum (pp). Sow responsiveness towards Crush and Handling Test was tested on Day 3 pp, responsiveness towards real crushing was observed first 72 h pp. Parity and litter size were included in the models as fixed effects. Multiparous sows were scored as responders when they changed their posture from lying to sitting or standing. Data were analyzed by logistic regression (proc Genmod) in SAS. Sows in pens compared to crates showed a tendency to respond with higher probability on Crush Test (0.78 vs. 0.37, $\chi^2_{(1)}$=3.81; $P<0.1$), however responsiveness towards real crushing (0.64 and 0.6, $\chi^2_{(1)}$=0.4; ns) and Handling Test (0.26 and 0.28, $\chi^2_{(1)}$=0.01; ns) did not differ. Parity and litter size were non-significant in these models. Mortality did not differ in both systems (9.2% vs. 6.3%; $\chi^2_{(1)}$=0.59; ns). In conclusion, similar level of sow responsiveness towards piglet distress calls and mortality at weaning in both systems suggest that maternal behaviour might be effective in pens in terms of preventing piglet crushing. Sows in pens, in spite of having a possibility to move freely, did not have necessarily higher responsiveness towards piglet handling by a stock-person.

Effect of different free farrowing environments on lying behaviour of sows

Baumgartner, Johannes, Mueller, Carolin and Koller, Mario, University of Veterinary Medicine Vienna, Institute of Animal Husbandry and Animal Welfare, Veterinaeplatz 1, 1210 Vienna, Austria; johannes.baumgartner@vetmeduni.ac.at

In a free farrowing environment sows are allowed to move around and choose lying place freely. Both lying comfort and risk of piglet crushing are affected by pen design. Three commercial free farrowing systems were analysed with regard to lying behaviour of sows. FS1: 7.6 m^2, solid lying area, minimal straw, slatted dunging area; FS2: 4.9 m^2, fully slatted, central anti-crushing bars; FS3: 4.1 m^2, fully slatted, trapezoid shape. 68 sows (n=20/23/25) were observed continuously via video on four days. Lying behaviour was analysed for frequency, duration, position and localization. Additionally video recordings were analysed for crushing situations of piglets. Data were analysed using descriptive statistics. Lying was the predominant behaviour of sows in all systems (63-78%). Highest activity was found at the day before birth and lowest the day after birth irrespective of pen type. The amount of lying behaviour was highest in FS3. In FS1 the separated lying area was intensively used for lying (92% of lying time). Sows in FS2 mostly had to lie between the central bars (63%). FS3 sows had contact to the pen walls for 63% of lying time and had to rest in a cramped position with and intensive contact to pen equipment for 54 minutes per day. 51/44/61 crushing situations were observed of which 37%/27%/13% occurred during birth and in the first day of life (37%/54%/49%). Sow's changes of the lying posture were most dangerous for the piglets to get crushed (54%/50%/72%), the centre of the pen was the most critical pen sector. The high level of activity immediately before parturition indicates the sow's need for nest building activity. We suppose that the unhindered area for lying was too small in each farrowing pen investigated. Enlargement of the lying area should help to improve lying comfort of sows and to minimise piglet crushing.

Outdoor and indoor rearing effects on behaviour and performance of sows and piglets

Juskiene, Violeta, Juska, Remigijus, Leikus, Raimondas and Razmaite, Violeta, Lithuanian University of Health Sciences, Institute of Animal Science, R. Zebenkos 12, 82317, Baisogala, Lithuania; violeta@lgi.lt

The purpose of the study was to compare the behavior and performance of sows and piglets raised outdoors and indoors. The study was carried out during summer period from May till August. Sows (n=12) of Lithuanian native breed were allotted into two groups analogues by origin, age and weight. Sows kept indoors (Indoor group) two weeks before farrowing were driven to the farrowing place and kept individually in pens of 8.7 m^2 area. Sows kept outdoors (Outdoor group) at month 4 of pregnancy were driven into the outdoor enclosures designed for 3 sows. Three wooden kennels (of 9.1 m^2 area) were fixed in each enclosure for each sow. The growth intensity of pigs was determined by individual weighing. Behaviour of sows and piglets were observed by one's eyes on three occasions per group every 15 minutes during one day per week from 8.00 a.m. to 8.00 p.m. The behaviour observations were recorded from the begining of study until 21 day of piglets' age. Statistical analysis of the data was performed using MS Excel operating system. The difference was considered significant when $P<0.05$. The study indicated that average weight of piglets in Outdoor group at the age of 3 weeks was 21.9% higher ($P<0.05$) than that of piglets in Indoor group, but survival rate until this age was 9.3% lower. The behavioural studies indicated that before farrowing sows in Outdoor group choose to rest all together in one kennel and spent 61.4% of time laying together. After farrowing each sow with it‘s piglets lay in separate kennels. No conflicts and agressiveness were observed among sows during behaviour studies. Sows kept indoors at weeks 2 and 3 after farrowing used to lay 6.5-27% more, while sows kept outdoors correspondingly used to be 10.9-25.7% more active. Piglets born and raised outdoors at week 2 and 3 used to lie, respectively, 11.9 and 11.1% less and used to be 12.9 and 16.9% more active than those born and kept indoors. The results of this study indicated that outdoor keeping of sows and piglets have positive impact on their activity and growth rate of piglets.

Effects of an alternative farrowing pen on performances of lactating sows and their litters

Jeon, Jung-Hwan[1], Song, Jun-Ik[1], Lee, Jun-Yeob[1] and Kim, Du-Hwan[2], [1]National Institute of Animal Science, Animal Environment, 77 Chuksan-gil, Kwonsun-gu, Suwon 441-706, Korea, South, [2]Gyeongnam National University of Science and Technology, Animal resources Technology, 33 Dongjin-ro, Jinju 660-758, Korea, South; jeon75@korea.kr

Crushing is closely related to sow behaviour and often the death of suckling pigs is caused by crushing. Hence, farrowing crates are generally used in the swine industry to reduce the crush rate which farrowing crates with support bars prevent sows from suddenly rolling or lying down. However this farrowing crate restricts sow movement to protect piglets from being crushed by sows. Moreover, animal welfare is becoming more important in the livestock industry. 64 sows were randomly divided into two groups at 7 days before the expected farrowing date. The alternative farrowing pen had support bars to prevent sows from suddenly rolling or lying down which is a variable crate (installed until 5 d after birth). Piglets were cross-fostered before they were 24 h old, and were supplied creep feed conventionally after ten days of age. Sows were fed a standard ration of commercial concentrate twice a day (07:00~08:00 and 18:00~19:00). They had free access to water at all times. Chi-square analysis was used to determine significant differences in crushed piglets and estrus interval. There are no difference in feed intake and back-fat thickness ($P>0.05$). The estrus interval was shorter in the alternative farrowing pen (treatment group; 5.6±0.5 d) than in the farrowing pen (control group; 6.5±1.0 d) ($P=0.0184$). Also, birth weight, weaning weight, and crushed piglets did not differ between the control and the treatment group ($P>0.05$). We found that an alternative farrowing pen did not affect the crushing rate, whereas it did improve the estrus interval. Based on these results, an alternative farrowing pen is expected to improve sows' activity without influencing the performance of sows and their litters. We therefore suggest that this housing may improve the lactating sow's welfare without supplementary area.

Improving quality of life for both sows and piglets remains a challenge during farrowing and lactation

Moustsen, Vivi A.[1]*, Pedersen, Janni H.*[2] *and Hansen, Christian F.*[2]*,* [1]*Danish Pig Research Centre, Agricultural and Food Council, Housing and Environment, Vinkelvej 11, 8620 Kjellerup, Denmark,* [2]*Faculty of Health and Medical Sciences, University of Copenhagen, HERD - Centre for Herd-oriented, Education, Research and Development, Grønnegårdsvej 2, 1870 Frederiksberg C, Denmark; hales@life.ku.dk*

Designing farrowing pen in order to maximise welfare for sows and piglets and at the same time remain competitive in a global pork market remains a challenge. Based on research in pig welfare and behaviour a number of pens for loose farrowing sows were implemented in three commercial production herds, so the herds had both crates and pens. During a two year period a total of 530 farrowings in loose pens (L) and 329 farrowings in crates (C) were recorded. Mean (±SE) sow parity was 3.4±0.06 and mean (±SE) number of total born piglets was 17.0±0.12 piglets/litter in both systems. Data were analysed using logistic regression with herd, littersize and pen as fixed effects for 'dead from birth to litter equalisation' and 'dead from litter equalisation until weaning of number of piglets in a equalised litter. The results showed increased total piglet mortality for L-sows compared to C-sows (3.9±0.09 vs. 3.2±1.12 piglets/litter, P<0.001). The main difference between L and C sows was that 44.5% of the L-sows vs. only 22.5% of the C-sows had two or more dead piglets in their litter between litter equalisation and weaning. In two of the herds, existing barns were refurnished. The pens were larger (5.3-6.3 m^2) than the crates they replaced, but compromises regarding pen dimensions and layout might have influenced function and productivity negatively in the pens. Using the same footprint for crates and pens might have negative impact on pen function and on productivity. The results indicate that loose farrowing is not yet robust under commercial situations. The pens have the capability to deliver the same performance but for a significant number of sows this is currently not the case and our challenge is to develop a system which will.

Effect of housing systems on horse behaviour: a possibility to keep stallions in group

Briefer Freymond, Sabrina[1], Von Niederhäusern, Ruedi[1], Bachmann-Rieder, Iris[1] and Briefer, Elodie[2], [1]Swiss National Stud Farm, Les Longs Prés, P.O. Box 191, 1580 Avenches, Switzerland, [2]Queen Mary University of London, Biological and Experimental Psychology Group, Mile End Road, London E1 4NS, United Kingdom; sabrina.briefer@haras.admin.ch

Horses are rarely kept in groups, despite such housing system fulfilling most of their welfare needs and mimicking bachelor bands that are found in the wild. Unfortunately, the high level of aggression that unfamiliar stallions display during encounters discourages owners to keep them in groups, because of the potential risks of injuries. Here, we investigated the welfare benefits and feasibility of group housing in breeding stallions. We integrated 5 stallions in a group on a pasture (4 hectares) in 2009 and 8 stallions in 2010 (8- to 19-year-old Franches-Montagnes). We compared the daily time budget of various behaviours (e.g. feeding, resting, locomotion) displayed when stallions were in individual stables versus outdoor groups. Then, we observed the quantity (occurrence) and quality (positive/negative) of interactions occurring during the first 21 days after group integration. Finally, we investigated the link between the final dominance ranks of males, assessed using pair feeding tests, and the occurrence of social interactions during the integration phase. We found that the housing system significantly affected various behaviours and their time budget. Horse were standing stare more often and tended to have more interactions in pasture than in individual stables (ANOVA: stand stare, $F_{1,52}$=9.06, P=0.004; interactions $F_{1,52}$=4.02, P=0.05). Negative interactions decreased quickly during the first four days after integration (GLMM: t_{1224}=-4.88, P<0.0001). In contrast, positive interactions increased over the integration period (t_{1226}=-3.29, P=0.001). More dominant males had less positive interactions (t_6=-2.73, P=0.034) and more negative interactions than lower-ranking males (t_6=-3.10, P=0.021). These results indicate that housing breeding stallions together is feasible in these conditions, because the number of negative interactions decreases and is kept at a minimum after only four days following group integration. Therefore, this housing system seems not to be associated with high risks of injuries, increases horse welfare and reduces economic costs linked to horse management.

Effects of moderate weather conditions on shelter behavior of horses

Snoeks, Melissa[1], Bax, Marijke[2], Permentier, Liesbet[1], Driessen, Bert[2] and Geers, Rony[1], [1]KU Leuven, Department of Biosystems, Bijzondere Weg 12, 3360 Lovenjoel, Belgium, [2]K.H.Kempen, Dier&Welzijn, Kleinhoefstraat 4, 2440 Geel, Belgium; Melissa.Snoeks@biw.kuleuven.be

The effect of weather parameters on the sheltering behavior of horses at pasture was studied. A first study with 40 horses included 244 observations (instantaneous sampling) spread over summer and autumn. The effects of weather parameters on the sheltering behavior (mean±SEM; 12.44±2.11%) were analyzed by using chi-square tests and linear mixed models (MIXED, CHISQ; P<0.05) with season (summer, autumn), presence or absence of sunshine, rainfall, cloudiness, dry air temperature, wind speed, relative humidity (RH) and all 2-way interactions as fixed effects and horse as random variable. Sheltering behavior was defined by standing under or standing within a horse length near artificial or natural shelter. Sheltering behavior was least at: a combination of a moderate wind speed and a RH between 70 and 80% (P<0.0001), temperatures above 25 °C (P<0.0001), a moderate wind speed (P=0.0055) and during summer (P<0.0001). The interaction between temperature and RH (P<0.001) and between temperature and wind speed (P<0.0001) was significant. Rainfall, RH or temperature in combination with wind speed had no significant effect. The second study, with observations during summer, autumn and winter, focused on the preference of horses when providing artificial or natural shelter. Natural shelter was mostly provided by trees in or next to the pasture. The artificial shelter was different in each pasture: a pasture with three individual stables, one with a lean-to, a semi-open stable, a classical shelter and an old truck. The time spent under artificial shelter was shorter (MIXED; P<0.0001) than under natural shelter. The sheltering duration under natural shelter was positively influenced (WILCOXON) by average RH (P=0.001), freezing weather (P=0.002) and by none or lightly clouded weather (P=0.002) but not by wind direction, wind speed, sunshine or rainfall. Sheltering duration under artificial shelter was negatively influenced by freezing weather (P=0.039).

Cattle behavioural and physiological responses to the amount of shade in pasture-based dairy systems

Schütz, Karin E.[1], Cox, Neil R.[2] and Tucker, Cassandra B.[3], [1]AgResearch Ltd, Ruakura Research Centre, East street, Private Bag 3123, Hamilton 3240, New Zealand, [2]AgResearch Ltd, Invermay Research Centre, Puddle Alley, Private Bag 50034, Mosgiel, New Zealand, [3]University of California, Department of Animal Science, 1 Shields Ave, Davis, CA 95616, USA; karin.schutz@agresearch.co.nz

Access to shade reduces heat load in cattle and improves production and welfare. We investigated the effect of the amount of shade on cattle behaviour and physiology. Eight Holstein-Friesian herds were studied for two consecutive summers (mean temperature:23 °C, range:17-29 °C) on six commercial, pasture-based dairy farms. Farms were selected based on variation in the amount of natural shade provided (all farms had unshaded and shaded paddocks). The number of cows in shade, near water (≤10 m of the trough), and with panting scores ≥2 [0 (no panting) to 3.5 (high respiration rate, excessive drooling, tongue extended)] were recorded every 30 min between 10:00-15:30 in shaded (n=31, range:0.1-14.3 m^2 shade/cow) and in unshaded (n=11) paddocks. Data were analysed using REML. All cows in a herd were physically able to use the shade simultaneously when the shade amount exceeded 2 m^2/cow. Herds in paddocks with 2-5 m^2 shade/cow used the shade less (max% cows in shade averaged 44%; range:13-100%, 2 of 11 observation days exceeding 80%) than those with shade amount exceeding 5 m^2 (average:80%, range:44-100%, 8 of 11 days exceeding 80%, P=0.004). A higher percentage of cows with 2-5 m^2 shade/cow tended to have panting scores ≥2 (3.5% of herd, range:0-10%) than those with 5 m^2 or more shade (1.6% of herd, range:0-5%, P=0.118). Max% of cows in shade increased by 3.1% for every m^2 increase in shade size (SE=1.51, P=0.042). Despite these differences in shade use, the maximum proportion of the herd near water (P=0.728) or with panting scores ≥2 (P=0.162) was unaffected by shade size. When there was no shade, a larger proportion of the herd was near the water trough as the weather became warmer (2.7% more animals/°C, SE=0.77%, P=0.008) and had panting scores ≥2 (3.5 animals/°C, SE=1.56, P=0.054). The results indicate that providing more shade allows a higher proportion of animals to simultaneously use this resource.

Providing artificial shade for beef cattle kept in feedlots increases time at the feed trough and improves weight gain

Macitelli Benez, Fernanda[1], Ribeiro Soares, Désirée[1], Rodrigues Paranhos Da Costa, Mateus J.[1], Marion, Pamella[2] and Cardoso Romano, Fernanda[2], [1]São Paulo State University, Animal Science, Via de Acesso Prof. Paulo Donato Castellane, 14884-900, Brazil, [2]Federal University of Mato Grosso, Animal Science, Rodovia Rondonópolis-Guiratinga, KM 06, 78735-910, Brazil; fmacitelli@yahoo.com.br

The aim was to evaluate the effect of artificial shade on average daily gain (ADG) and time spent at the feed trough (TST) in beef cattle feedlots. 120 Nellore bulls (24±4 mo.) were assigned during 63 d in two treatments: (T1) with shade access or (T2) no-shade access (n=3 groups/treatment; 20 animals/group). Shade was provided whole day (4 m^2/animal) by wooden structures covered with aluminum shading (blocked 80% of solar radiation), installed in the back half of the pen. Feedlot pens (11 m^2/animal) had feed bunk with 0.55 m/animal. Diet was the same for all treatments (Mombaça grass silage and concentrate) delivered four times/day and TST was monitored twice a week (d1 to d15) and once a week (d16 to d63) using instantaneous scan sampling conducted at 10 min intervals (07:30 h to 17:30 h). ADG was calculated by period: P1- d1 to d30, P2- d31 to d63, Pt- d1 to d63. Means (±sd) ambient temperature and relative humidity were 34.80±5.85 °C and 30.17±15.11%, respectively. ADG and TST were compared using general linear model that included treatment as fixed effects to ADG and the same effect plus period and treatment × period to TST. Animals of T1 had higher ADG than T2 (2.35±0.62 kg/day vs. 1.84±0.53 kg/day; $P<0.05$) at P1; but not at P2 (1.75±0.51 kg/day vs. 1.81±0.52 kg/day; $P>0.05$) nor Pt (1.99±0.31 kg/day vs. 1.89±0.34 kg/day; $P>0.05$). TST was the same at P1 (85.48±20.12 min/day vs. 85.18±23.82 min/day; $P>0.05$), but less at P2 (86.99±18.95 min/day vs. 99.50±25.33 min/day; $P<0.05$) and Pt (86.24±20.64 min/day vs. 92.34±24.69 min/day; $P<0.05$). We conclude that animals with shade access had better adaptation to feedlot shown by higher performance at the beginning of feedlot with the same time at the trough. Moreover, the same performance with less time at the feed bunk in the end of fattening period.

Bulls benefit from rubber-coated slats

Härtel, Heidi[1], Lappalainen, Marko[2], Wahlroos, Taru[3], Piipponen, Maria[3] and Hänninen, Laura[3,4], [1]HKScan Finland, HK Agri, Teollisuuskatu 17, 30420 Forssa, Finland, [2]Nautakasvattamot.fi, Lentokentänkatu 5b, 33900 Tampere, Finland, [3]University of Helsinki, Department of production animal medicine, P.O. Box 57, 00014, Finland, [4]University of Helsinki, Research centre for animal welfare, P.O. Box 57, 00014 Finland; laura.hanninen@helsinki.fi

In Northern latitudes bulls are commonly kept in groups on concrete slatted floors, but they might benefit from rubber-coated concrete slats. We observed bulls´ resting behaviour (lying-down duration and style) and scored front-knee and hock lesions (from 0: no lesions to 2: severe lesions) for 12 months in 2 month intervals. The study was conducted on a private farm, with 6 pens of 20 bulls. Bulls were moved to barn at 6 month of age, and were slaughtered at 18-20 months (app. 355 kg slaughter weight). Half of the pens had two areas: rubber coated slats (ANIMAT) and concrete slats (COATED). The identical control pens had only original concrete slat-floors (CONCRETE). The depths of the pens were always 6 m, and the space allowances varied within age: 2 m^2 (6-10 months), 2.4 m^2 (10-14 months), and 3 m^2 (>14 months of age). The available coated-slats/animal were respectively: 1.2 m^2, 1.4 m^2 and 1.8 m^2. The effects of the floor type on the front-knee and hock lesions were analyzed with a linear mixed model taking repeated observations into account. The effect of floor type on the overall median lying-down time and style were analyzed with Mann-Whitney-U-tests. The median (min-max) lying-down duration was longer in CONCRETE pens than in the COATED pens (P=0.009): 7.2 s (7.1-9.3 s) vs. 4.8 s (3.2-4.0 s). The occurrence of abnormal lying-downs (P=0.002) was higher in the CONCRETE pens than COATED ones: 37% (0-71%) vs. 0% (0-25%). CONCRETE bulls had overall more severe leg lesions than COATED bulls (P<0.01 for both). Bulls resting behaviour and leg health was improved in pens with rubber-coated resting areas compared with bulls kept in pens with fully concrete-slatted floors. However, results are only from one farm, and further studies are needed.

Indirect but not direct encounters with manure scrapers induce signs of stress in cows

Gygax, Lorenz[1], Buck, Melanie[1], Wechsler, Beat[1], Friedli, Katharina[1] and Steiner, Adrian[2], [1]Swiss Federal Veterinary Office, Centre for Proper Housing of Ruminants and Pigs, Tänikon, 8356 Ettenhausen, Switzerland, [2]Vetsuisse-Faculty of Berne, Clinic for Ruminants, Department of Clinical Veterinary Medicine, Bremgartenstrasse 109a, P.O. Box 8466, 3001 Berne, Switzerland; lorenz.gygax@art.admin.ch

Cows are frequently faced with automatic manure scrapers in loose housing but whether such encounters impact their welfare remains unknown. We observed behaviour of 29 focal lactating cows during ten manure scraping events on three Swiss working farms while simultaneously recording mean intervals between heart beats (RR-intervals; ms) and heart rate variability (square root of the mean squared difference of successive RR intervals, RMSSD; ms) in 30s time windows. Additional reference values were collected in periods without scraping event. Data were analyzed using linear mixed effects models. RMSSD values were 18% lower (95% CI=[2;32]; $F_{1,130}$=4.84, P=0.03) when cows were indirectly faced with a scraping event compared to situations with no scraping i.e. when they were either lying, standing/walking, standing in the cubicle with all four legs or standing in the cubicle with the front legs only. Comparing direct encounters to standing/walking in the alley without a scraping event, cows tended to have lower RMSSD values when singly avoiding the scraper (-40% [-61;-9]), but not when doing so in a crowd (-21% [-59;+52]), nor when standing/walking in the alley during scraping events (-37% [-55;-11]), crossing the scraper without (-11% [-43;+41]), nor with touching (-21% [-50;+25]; $F_{5,75}$=2.19, P=0.06). Similarly, we did not find a difference in RMSSD values between situations when cows were feeding without scraping event and when they directly encountered the scraper during feeding (-10% [-35;+25]) or when being displaced by the scraper from feeding (+30% [-16;+101]; $F_{2,36}$=1.51, P=0.24). There was no statistically detectable difference in mean RR-intervals in any of these comparisons. Given the effects of scraping events on RMSSD, it seems that cows indirectly faced with a scraping event experience some signs of stress whereas no such reaction was detectable during direct encounters.

Effect of construction geotextile mattress on resting behaviour of cows

Uhrinčať, Michal[1], Hanus, Anton[1], Tančin, Vladimír[1,2] and Brouček, Jan[1], [1]Animal Production Research Centre Nitra, Hlohovecká 2, 951 41 Lužianky, Slovakia (Slovak Republic), [2]Slovak University of Agriculture, Trieda A. Hlinku 2, 949 76 Nitra, Slovakia (Slovak Republic); uhrincat@cvzv.sk

The straw price is higher every year and farmers are forced to seek alternatives to ensure the welfare of their animals while resting in free-stall housing. One of the alternatives is mattresses made from different materials. We hypothesized that these materials affect the resting time. Twenty-four non-lactating, dry Holstein dairy cows without experience with mattresses were housed individually in separate pens. Each pen (345×485 cm) contained 3 free stalls 115 cm wide, 235 cm long with a 65 cm front lunge and different mattress 170 cm long: the rubber crumb filled mattress with bigger (M1) (3-8 mm) or smaller (M2) (1-4 mm) particles and mattress (M3) of pressed PUR segments. The stalls where covered by geotextile. For the first three days the animals were allowed access to all three stalls for acclimatization. During restriction phase – the next 9 days, the animals were allowed access to only 1 of the 3 stalls at a time (in a random order for the first stall) for each additional three days. Behaviour was video-recorded, last 24 hours in 1 minute sampling of each interval was evaluated. Differences between mattresses were analysed using a mixed model (SAS 9.1). The average lying time was the highest in the M3 mattress (13.2±0.6 h) (mean±SE), but we observed no significant differences in relation to mattresses M1 or M2 (12.5±0.7 h; 12.7±0.7 h resp.). Time spent lying on left side and the number of lying bouts was significantly higher ($P \leq 0.05$) in M3 mattress 6.9±0.4 h lying and 26.5±2.4 bouts compared to M1 mattress 5.8±0.5 h lying and 21.9±1.7 bouts. These results indicate that assessed mattresses are comparable in terms of lying time, but cows during the last gestation months prefer the left lying side, thus the mattress M3 could be seen as more suitable.

The effect of lying deprivation on cow behavior

Norring, Marianna[1], Valros, Anna[1] and Munksgaard, Lene[2], [1]Research Centre for Animal Welfare, Faculty of Veterinary Medicine, P.O. Box 57, 00014, University of Helsinki, Finland, [2]Aarhus University, Faculty of Agricultural Sciences, P.O. Box 50, 8830 Tjele, Denmark; marianna.norring@helsinki.fi

In practice cows may spend time standing waiting for access to milking, feed and cubicles. It has been suggested that high yielding cows may be in a trade off situation between eating and lying especially in systems with long waiting time before milking. Therefore, we tested the effect of preventing cows with different production levels from lying down on their lying motivation. We used 32 cows of parities 1-7 in tie-stalls. The cows entered the experiment at 8 weeks post partum. They were milked at 6:00 and 18:00. The behavior of the cows was observed during 2 control days followed by 2 treatment days when the cows were deprived of lying either from 10-14 or from 14-18. Lying, eating, ruminating and lying with neck muscles relaxed (used as an indicator behavior for sleep) were observed from 14-18 and 18-22, and milk yield was recorded. The percentage of time allocated to different behaviors was analyzed using mixed models. When the behavior after lying deprivation was compared to control days we found that the cows were lying down more ($F_{1,30}$=18.9, P<0.001; control 24% vs. lying deprived (10-14) 35%). The cows spent less time eating after forced standing (14-18) ($F_{1,30}$=6.4, P=0.017; control 24% vs. lying deprived (10-14) 21%). Cows used more time lying inactive without ruminating and lying neck relaxed post treatment compared to control days ($F_{1,31}$=13.2, P=0.001; control 12% vs. lying deprived (10-14) 20%). During the period 18-22 lying deprived cows were lying more (P=0.031) and eating less (P=0.032) compared to control days. Yield had no effect on post deprivation behavior. In conclusion, on short-term lying and sleeping behaviors increase following deprivation indicating increase in motivation however, milk yield had no effect on this motivation.

The effect of feeding regime and presence of mother on rumination in young dairy calves

Ellingsen, Kristian[1], Mejdell, Cecilie M.[1], Westgård, Silje[2], Dangstorp, Gunhild[2], Johnsen, Julie[1] and Grøndahl, Ann Margaret[1], [1]Norwegian Veterinary Institute, Section for disease prevention and animal welfare, P.O. Box 750 Sentrum, 0106 Oslo, Norway, [2]Kalnes Agricultural College, Sandtangen 85, 1712 Grålum, Norway; Kristian.Ellingsen@vetinst.no

In calves it is believed that a restrictive feeding regime of approximately six litres of milk per day motivates for roughage and concentrate intake at an earlier age, and promotes development of rumen function. A Swedish study, however, found that calves suckling freely ruminated more at two weeks of age than calves fed nine litres of milk substitute per day from an automatic feeder (AF). The aim of this study was therefore to investigate the effect of presence of mother and level of satiation on the development of rumination in young calves. Norwegian Red calves (n=24) were allocated into four groups: (1) separated immediately after birth and fed six litres of milk/day from AF (n=6, group pen); (2) separated immediately after birth, fed ad lib milk rations from AF (n=6, group pen); (3) kept together with the mother for the full duration of the experiment (22 days) (n=6, individual calving pen); (4) kept with the mother for three days (n=6, individual calving pen) and thereafter given the same treatment as calves in group one. The calves were observed for two hours at 10, 12, 14, 16, 18, 20 and 22 days of age. At day 10, four of 23 calves were observed ruminating. At days 14 and 20 the proportion of calves that had been observed ruminating was 11 and 18 of 23, respectively. T-tests showed that group 4 spent significantly more time ruminating at three weeks of age ($P<0.05$) than groups 1, 2 and 3. This may indicate that the dam has an effect on development of rumen function but that calves which are satiated on milk are less motivated to eat solid feed at such a young age. It may also indicate that the amount of milk fed to calves through AFs has no effect on ruminal development.

Effect of early assistance at calving on lying behaviour of dairy cows and calves

Villettaz Robichaud, Marianne[1], Haley, Derek[1], Jeffrey, Rushen[2], Godden, Sandra[3], Pearl, David[1] and Leblanc, Stephen[1], [1]Ontario Veterinary College, University of Guelph, Population Medicine, 2509 Stewart Building (#45), Guelph, ON, N1G 2W1, Canada, [2]Agriculture and Agri-Food Canada Research Centre, 6947 # 7 Highway, P.O. Box 1000, Agassiz, BC, V0M 1A0, Canada, [3]College of Veterinary Medicine, University of Minnesota, 225 Veterinary Medical Center, 1365 Gortner Avenue, St. Paul, MN 55108, USA; marianne.villettaz@gmail.com

Calving events are crucial times in a dairy cow's life. It is around this time that the cows and calves are at higher risk of sickness and death. Research is needed to investigate management practices that could help reduce those risks. Previous research has suggested that lying behaviour could be used as an indicator of discomfort in dairy cattle. The aim of this study was to evaluate the impact of early assistance at calving on lying behaviour in dairy cows and calves. We measured lying behaviour for 5 days post calving for 20 newborn female Holstein calves and 30 Holstein dairy cows using validated accelerometers (Hobo data loggers). Half the cows (n=15) were assigned at random to early assistance, defined as pulling the calf approximately 15 minutes after first sight of both hooves. The other 15 cows did not receive assistance at calving. The calves were also equally distributed between assisted and unassisted calving. There was no difference between the assisted and unassisted groups in the number of lying bouts over 5 days for the cows (mean/day ± SD: 11.6±4 vs. 13±13.8 or the calves (20.8±4.4 vs. 22±4.2; T test: P=0.30). However, on day 1 post calving, unassisted cows tended to have more lying bouts than assisted cows (14±5 vs. 11±5; T test: P=0.10). The total hours spent lying down over 5 days post calving tended to be higher in unassisted cows compared to assisted cows (55.8±12.8 h vs. 48.0±10.8 h; T test: P=0.11) and higher in assisted calves compared to unassisted calves (101.8±4.7 h vs. 98.7±3.0 h; T test: P=0.11). These results suggest that early assistance at calving applied by pulling can affect lying time of dairy cows and calves in the days following calving.

Head partitions at the feed barrier affect social behaviour of goats

Nordmann, Eva[1], Barth, Kerstin[2], Futschik, Andreas[3] and Waiblinger, Susanne[1], [1]University of Veterinary Medicine, Institute of Animal Husbandry and Welfare, Veterinärplatz 1, 1210 Vienna, Austria, [2]Johann Heinrich von Thünen – Institute (vTI), Institute of Organic Farming, Trenthorst 32, 23847 Westerau, Germany, [3]University of Vienna, Department of Statistics and Operations Research, Universitätsstrasse 5/9, 1010 Vienna, Austria; eva.nordmann@vetmeduni.ac.at

Space allowance at the feeding places often forces goats to feed in close proximity, that is, less than their individual distances. In consequence, agonistic behaviour may increase as well as stress and injuries. The aim of this study was to investigate the influence of non-transparent head partitions at the feed barrier on agonistic behaviour in loose-housed horned dairy goats. The study involved two groups of 36 animals each in pens with a wooden palisade feed barrier (one feeding place/animal). Hay was fed *ad libitum* and provided at 08:30. At 13:00 and 17:30, the remaining hay was shuffled closer. Each group was tested with and without non-transparent head partitions for 11 experimental days per treatment. Agonistic interactions were recorded for 13.3 h per group and treatment (2.7 h on 5 days/group distributed over the main feeding times (08:30-10:30, 13:00-15:00, 17:30-18:50)). Dominance values were used to allocate goats to rank categories (0.00-0.33=low-ranking (n=23), 0.34-0.66=middle-ranking (n=27), 0.67-1.00=high-ranking goats (n=22)). To test if goats more often feed in adjacent feeding places with head partitions, the number of changes between an occupied and an unoccupied feeding place (Changes_Occupancy) were observed via scan sampling every 10 min during 48 h per treatment. Data were analysed by Wilcoxon-tests or T-Test (Changes_Occupancy). With head partitions a lower number of displacements from feeding place by an actor standing inside the feed barrier was found (median=2.0 interactions/goat × 13.3 h, min-max=0.0-12.0; without: 2.9, 0.0-12.9; P=0.002). The impact was most pronounced in low-ranking animals (P=0.009), but effects were also found in middle-ranking goats (P=0.030). With head partitions more goats were feeding directly next to each other (Changes_Occupancy lower, P=0.017). In summary, non-transparent head partitions seemed to reduce the accepted distance between goats and therefore showed beneficial effects in terms of lower levels of social disturbances during feeding. We acknowledge funding by BMLFUW and BMG, Project-Nr.100191.

Early adaptation of lambs to a feedlot system

Jongman, Ellen[1], Butler, Kym[1], Rice, Maxine [2] and Hemsworth, Paul[2], [1]Animal Welfare Science Centre, Department of Primary Industries, FFSR, 600 Sneydes Road, Werribee 3030, Australia, [2]Animal Welfare Science Centre, University of Melbourne, Parkville, Parkville 3049, Australia; Ellen.Jongman@dpi.vic.gov.au

The environment in which we keep domestic livestock requires behavioural adaptation in these animals. Although extensive pasture-based production of lambs remains the dominant system for finishing lambs in Australia, the use of feedlot systems is increasing. This study examined the relationships between social and feeding behaviour and stress in lambs held in feedlots with the aim to refine measurements to be taken during future studies. Results of the first week (indication of early adaptation to the feedlot environment) are presented. Lambs in two 20-lamb feedlots (with 2 m^2 floor space/animal) were studied. Prior to entry to the feedlot the lambs were subjected to a temperament test (Isolation Box Test). Weight gain, general activity, lying time, time at the feeder (recorded via data logging devices), ability to displace other lambs at the feeder (video observation during peak feeding time) and basal plasma cortisol (1 sample using venepuncture) were recorded. A parsimonious general linear model was developed to relate the logarithm of cortisol concentration to all other measurements. The parsimonious model accounted for 36% of the variance in cortisol (P=0.0005), with predicted cortisol values differing about 10 fold within the range of behaviour data observed. Cortisol was most elevated if the lamb's activity in the temperament test (number of steps) was high but its activity (number of steps) in the feedlot was low. Cortisol was also high if the lamb's activity in the feedlot was low and it was displaced by other lambs at the feeder. The results indicate that the stress response (cortisol) during early adaptation to a feedlot is related to the interaction of individual temperament and behaviour restrictions imposed by the environment on individuals. The parsimonious general linear model was able to disentangle temperament and direct behavioural responses on cortisol that were difficult to detect when analysing individual correlations.

Lambs preferences for different types of bedding materials

Teixeira, Dayane L.[1], Miranda-De La Lama, Genaro[1], Villarroel, Morris[2], Escós, Juan[1] and María, Gustavo[1], [1]University of Zaragoza, Department of Animal Production and Food Science, Calle Miguel de Servet 177, 50013, Spain, [2]Polytechnic University of Madrid, Department of Animal Science, E.T.S.I.A., Calle de Ramiro de Maeztu 7, 28040, Spain; dadaylt@hotmail.com

We investigate lambs´ preference when four different flooring materials were available simultaneously. A total of sixteen entire male lambs of Rasa Aragonesa breed (21.12±0.86 kg, 75 d old) were fattened for 18 days. Two groups of eight lambs were housed in pens being divided into five areas. Four areas had one type of flooring: sawdust (SD), industrial waste of paper and pulp (WP), straw (ST) and rice husks (RH), while the fifth had cement floor (CT). A video recording device was set up to record maintenance behaviour by scan sampling every 10 minutes (07:00-21:00) throughout the experiment. The behaviour recorded included lying down, standing, walking, feeding and drinking. Kruskal Wallis testing was carried out on the frequency rates with which the animals performed each behaviour pattern in each of the different areas during the entire fattening period. Mann-Whitney U test was used to compare pairs, with penalty weighting. In general, the occupation rate was significantly different ($P \leq 0.001$). Lambs demonstrated preference for SD bed, with a 47% occupation rate, followed by WP (17%), ST and RH, with occupation rates of nearly 7%, respectively. The occupation in the area with cement floor was 21%. Significant differences were also found between lying down and standing behaviours in different bedding zones. Respectively, the percentages of these behaviours were: 80% × 20% on the SD bedding; 63% × 37% on the WP bedding; 23% × 77% on the ST bedding; and 40% × 60% on the RH bedding. The preference for sawdust may be associated to the sawdust greater absorption properties compared to straw. It is also significant that a bedding material that was completely new to the lambs as is WP should be their number two preference.

Effects of group size and floor space allowance on aggression and stress in grouped sows

Hemsworth, Paul H.[1], Rice, Maxine[1], Giri, Khageswor[2] and Morrison, Rebecca[3], [1]The Animal Welfare Science Centre, The University of Melbourne, Department of Agriculture and Food Systems, Parkville, 3010, Australia, [2]The Animal Welfare Science Centre, The Department of Primary Industries, Werribee, 3030, Australia, [3]Rivalea Australia, Corowa, 2646, Australia; phh@unimelb.edu.au

There is limited evidence that floor space and group size affect aggression and stress in grouped sows. We determined the effects of these two factors on aggression, stress, skin injuries and reproduction in grouped sows. A total of 3,120 sows were used in a factorial design with two main effects imposed within 7 days of insemination: group size of 10, 30 and 80 sows; and floor space allowance of 1.4, 1.8, 2.0, 2.2, 2.4 and 3.0 m^2 per sow. Sows were housed on partially slatted concrete floors and floor-fed 4 times per day. Aggressive behaviour (bites and knocks) at feeding was observed at days 2 and 8 of treatment, skin lesions (scratches, abrasions, cuts, abscesses, and partially healed or old lesions) were measured on days 2, 9, 23 and 51 and cortisol concentrations and neutrophil and lymphocyte counts were measured at days 2, 9 and 51. Venipuncture blood samples were collected from the jugular vein within 2 minutes of snaring to avoid an acute stress response to handling per se. Farrowing rate and litter size were also recorded. Each treatment was replicated 4 times over time. REML mixed model analyses examined treatment effect combinations after accounting for replicate and random spatial location effects within replicate. There was a consistent linear effect of space on aggression at feeding (day 2, P=0.029), plasma cortisol concentrations (day 2, P=0.0089) and farrowing rate (% inseminated sows that farrowed, P=0.012). There was a general decline in aggression and cortisol with increasing space and a general increase in farrowing rate with increasing space. Group size was related to skin lesions (day 9, P=0.00017; day 23, P=0.0046; and day 51, P=0.00059): groups of 10 had consistently low skin injuries. These results highlight the need to reduce aggression and stress in the immediate post-mixing period.

Effect of group size, phase of fattening period, gender and rank position on distances covered by fattening pigs

Brendle, Julia and Hoy, Steffen, Justus Liebig University, Department of Animal Breeding and Genetics, Bismarckstrasse 16, 35390 Giessen, Germany; julia.brendle@agrar.uni-giessen.de

Activity or locomotion in particular, defined as movement of an animal related to the change of location may lead to conclusions concerning health status, well-being and behavioural disorders of animals. In this study, distances covered by pigs under conventional farming conditions were analysed depending on different factors by using the software tool VideoMotionTracker®. The distances covered during 24 h depending on group size, phase of fattening, gender and social rank were studied in altogether 220 fattening pigs kept in groups of 6 or 12. Immediately after having grouped the pigs at the beginning of each fattening period the agonistic interactions of all group members were continuously recorded with infrared video technology during 72 h to analyse social hierarchy and the individual rank position of each pig. In three phases of the fattening period – at the beginning (Ø 80 days of age), in the middle (Ø 110 days of age) and at the end (Ø 140 days of age) – the distances covered by focus animals during 24 h were measured using the VideoMotionTracker® (Mangold). Statistical data analysis was carried out with the program package SPSS Version 19 applying the analysis of variance. Analysed traits were normally distributed. On average of all analysed pigs, the distances covered during 24 hours decreased ($P<0.001$) from 730 m at the beginning to 314 m at the end of the fattening period. Pigs in groups of 12 covered longer distances during 24 h compared to pigs kept in groups of 6 ($P<0.05$). Furthermore, lighter pigs covered longer distances than heavier pigs at the same age. Gender and rank position had no influence on covered distances. The study could prove that certain factors as mentioned above influence the distances covered by pigs. Results might lead to recommendations for the housing of pigs.

Supplementary feed offered to a tail biting pen changes feeding behaviour, feed intake, growth and tail health of fattening pigs

Palander, Pälvi[1], Valros, Anna[1], Heinonen, Mari[1] and Edwards, Sandra[2], [1]University of Helsinki, Department of Production Animal Medicine, P.O. Box 57, 00014 University of Helsinki, Finland, [2]University of Newcastle-upon-Tyne, School of Agriculture, Food & Rural Development, NE1 7RU, Newcastle-upon-Tyne, United Kingdom; palvi.palander@helsinki.fi

To study the capability and mechanism of supplementary feeds to stop tail biting, three different feeds were offered to fattening pigs on their pen floor for 14 days after tail biting was observed in 30 pens per treatment: feed with linseed, wheat bran and salts (LWS), linseed and wheat bran (LW) or only wheat bran (W). Changes in behaviour were measured from d 1 to d 15 at pen level in ten pens within each treatment. General activity was measured as body posture, eating, drinking and investigating the pen. Agonistic behaviours included tail biting (TB) and other agonistic (OA; ear biting, other social biting, threatening to bite and mounting). TB or OA were differentiated as in the feeder region (1 m^2) or elsewhere in the pen (15 m^2). Tail health and daily weight gain (DWG) were recorded on days 1, 3 (only tail health) and 15. Feed intake (FI) was recorded from two weeks before to two weeks after the treatment period. Behavioural changes were tested with one-way anova, tail health change and FI with repeated measures and DWG with univariate GLM. The proportion of scans eating at the feeder station decreased over time in LW and W (-0.9% and -0.3%, respectively) but increased in LWS (+1.1%, $P<0.05$). LWS had increased DWG compared to W and increased FI compared to LW ($P<0.05$). The proportion of healthy tails within the pen decreased over time, but to a lesser extent in LWS and LW than in W ($P<0.05$). Treatment had no effect on TB or OA, but 52±30% of all TB and 37±17% of all OA happened in the feeder region. Data suggests that nutritional content of supplementary feed had an effect on feeding behaviour, feed intake, growth and tail health of pigs.

Effects of enrichment material and feeding regime on exploratory behaviour of finishing pigs

Zwicker, Bettina[1], Gygax, Lorenz[1], Wechsler, Beat[1] and Weber, Roland[2], [1]Swiss Federal Veterinary Office, Centre for proper housing of ruminants and pigs, Tänikon, 8356 Ettenhausen, Switzerland, [2]Research Station Agroscope Reckenholz-Tänikon ART, Centre for proper housing of ruminants and pigs, Tänikon, 8356 Ettenhausen, Switzerland; bettina.zwicker@gmail.com

Provision of attractive enrichment material to pigs is crucial in intensive housing to meet their behavioural needs. This study aimed to investigate the effects of eight enrichment materials on exploratory behaviour of finishing pigs and their oral manipulation of pen-mates. In two experiments, 16 groups each of six pigs were housed in pens with partly-slatted floor. Half of the groups were fed either restrictively or *ad libitum*. Enrichment materials tested were cut straw, cut straw enriched with grains of maize, a straw block, and a rack filled with cut straw in Experiment 1, and chopped straw, chopped Miscanthus, bark compost, and a pellet dispenser in Experiment 2. Every three weeks, using a cross-over latin-square design, the enrichment materials were replaced. Behaviour was sampled by means of one-minute focal-scan sampling on the second and eighteenth day with a given material and analysed using linear mixed-effect models. In comparison to cut straw, the frequency of exploring was on average 28% higher with enriched cut straw (95% CI: [+10;+49]%) and lower with the straw rack (-15% [-27;-1]) and with the straw block (-75% [-78;-71]; $P<0.001$) and additionally decreased from the second to the eighteenth observation day (-19% [-27;-10]; $P<0.001$) in Experiment 1. In Experiment 2, there was an interaction between material and observation day ($P<0.001$). With all materials, pigs showed less exploratory behaviour over time, but this decrease was smaller for straw based materials. In both experiments, pigs fed *ad libitum* explored the materials only half as frequently as those fed restrictively (Experiment 1: -66% [-75;-55]; $P<0.001$). The frequency of manipulating pen-mates was not influenced by material or feeding regime in both experiments. It is concluded that enrichment materials vary in their long term value for pigs and that *ad libitum* fed pigs are less motivated to explore.

Chopped straw: a better nest building material for crated sows than newspaper or saw dust?

Valros, Anna, Swan, Kirsi, Yun, Jinhyeon, Mustakallio, Mikaela, Kothe, Stefan and Peltoniemi, Olli, University of Helsinki, Faculty of Veterinary Medicine, Department of Production Animal Medicine, P.O. Box 57, 00014 University of Helsinki, Finland; anna.valros@helsinki.fi

Several studies have shown benefits of facilitating sow nest building, but there is little data on the effect of different nest building materials. We provided crated sows with two handfuls of saw dust (SD) or chopped straw (CS), or two pages of newspaper (NP) twice daily prior to farrowing. Data was collected on farrowing duration, mean interval between piglet expulsions (INT), mean interval between the expulsion of piglets 3-4 and interval between the birth of piglets 1-2. The mortality cause of all dead piglets was defined by post mortem examination. Litter size, number of stillborn, total mortality of live-born (mortality) and total mortality ratio of live born (mortality ratio) were recorded. Treatments were compared with Kruskall Wallis tests and pairwise comparisons were done with Bonferroni-corrected Mann Whitney U-tests (PASW 18). All results are given as median and (interquartile range). There were no differences in farrowing duration between the treatment groups, even though INT was numerically highest in NP sows (21(17) min, n=14) compared to SD (18(14) min, n=10) and CS (15(9) min, n=12) sows. There was no difference in litter size or number of stillborn piglets between the treatments. Treatment affected mortality and mortality ratio significantly (P=0.02 for both). Pairwise comparisons revealed that this difference was mainly due to a higher mortality in NP (2(5) piglets, 15(21)%, n=19) than CS (1(2) piglets, 6(13)%, n=17) (P=0.01 and 0.02, respectively) litters. Piglet mortality in the SD sows (2(3) piglets, 12(24)%, n=23) was intermediate, with a tendency for a higher mortality ratio than the CS sows (P=0.09). These results indicate that the choice of nest building material is important, but the mechanism is still unclear.

Welfare assessment of pigs reared at pasture by monitoring enteric environment

Tozawa, Akitsu[1], Takahashi, Toshiyoshi[2] and Sato, Shusuke[1], [1]Tohoku University, Graduate School of Agricultural Science, 232-3, Yomogita, Naruko-Onsen, Osaki, Miyagi 989-6711, Japan, [2]Yamagata University, Faculty of Agricultural Science, 1-4-12 Kojirakawa-machi, Yamagata, Yamagata 990-8560, Japan; shusato@bios.tohoku.ac.jp

We postulated that pigs reared at pasture will change their enteric environment by eating plants and soils which will improve their welfare by improving their health. In this study, we used 44 castrated male pigs (LW) assigned to 3 groups. Group1 (G-1): pigs reared at pasture and fed fermented total mixed ration (n=14), Group 2 (G-2): pigs reared at pasture and fed concentrates (n=15), Group 3 (G-3): pigs reared in indoor pen with slatted floor and fed concentrates (n=15). We observed behaviour by 5 min. interval scan sampling from 9:30 to 16:30 for all groups (n=6 each). To assess the condition of enteric environment, we measured fecal pH three times (prior, middle, and latter period), and counted the numbers of all eubacteria and *Clostridium* cluster I, II from feces by real-time PCR (n=5 each). Clostridia include not only pathogenic bacteria but also species which produce skatole, so we measured backfat skatole in G-1 (n=5) and G-3 (n=4). Regarding the behaviour observations, time spent touching the soil (rooting and eating soil) was 14.5±9.2% in G-1 and 9.3±3.5% in G-2. Time spent eating and rooting, which are motivated internally in pigs, increased significantly in pasture rearing systems. Fecal pH was higher in the order of G-1 (6.6±0.2), G-2 (6.2±0.3), and G-3 (5.6±0.2). Percentages occupied by *Clostridium* cluster I, II in all eubacteria from feces were significantly higher in G-3 (12.4±6.4%) than in the other groups (G-1 : 4.2±3.8%, G-2 : 1.8±1.8%) (Kruskal Walis P=0.012). Backfat skatole was undetectable in 3 pigs out of 5 in G-1, in contrast, only 1 pig out of 4 in G-3. Rearing at pasture stimulated innate behaviour properly and reduced pathogenic intestinal flora (Clostridia) and skatole in backfat, which may lead to improved welfare in pigs.

Behaviour of fattening rabbits: comparison of housing in pens with and without straw bedding

Windschnurer, Ines, Meidinger, Katharina, Waiblinger, Susanne, Smajlhodzic, Fehim and Niebuhr, Knut, Institute of Animal Husbandry and Animal Welfare/University of Veterinary Medicine Vienna, Department for Farm Animals and Veterinary Public Health, Veterinärplatz 1, 1210 Vienna, Austria; Ines.Windschnurer@vetmeduni.ac.at

Intensively kept fattening rabbits are mainly housed in cages. This work evaluated the behaviour (e.g. agonistic, play, exploratory) and injuries of rabbits housed in pens as alternative housing system comparing straw bedding with slatted plastic floor. Over 3 fattening periods, 12 groups were observed on straw bedding and 12 on slatted floor (6 male and 6 female groups/system). 60 animals aged 38 days were placed in each pen. The night before slaughtering (aged 82 days) the behaviour of each group was analysed continuously using video-recordings (1 hour/pen, at 3am-4am). In the morning 30 rabbits/pen were examined for injuries caused by agonistic behaviour (score 0 [no] to 3 [deep injuries]). Potential system or gender effects on behaviour (frequencies/number of animals/pen) were evaluated by Mann-Whitney U Tests (pen as statistical unit). Injuries of individuals were analysed by Chi^2-tests. The two systems did not differ regarding the occurrence of agonistic behaviour, injuries, or play behaviour ($P>0.05$). Explorative behaviour (sniffing, licking, foraging, or scratching the wooden stick and other objects with forelimbs; without 'scratching the litter/floor') occurred more often on litter (mean±SD/animal*hour: litter: 1.90±0.447; slatted: 1.27±0.347; Z=-2.944, P=0.003). Rabbits on slatted floors scratched their conspecifics more often (litter: 0.04±0.042; slatted: 0.12±0.090; Z=-2.284, P=0.022). In contrast, rabbits on litter scratched the floor more often (litter: 0.11±0.121; slatted: 0±0.006; Z=-3.650, $P<0.001$). Male fattening rabbits showed more agonistic behaviour (male: 0.21±0.172; female: 0.05±0.066; Z=-2.691, P=0.007) and were injured more often than females (injured animals/examined: male: 42.1%, female: 15.5%; Chi^2=66.939, $P<0.001$). In conclusion, no floor effect was found on agonistic interactions, injuries, or play behaviour. However, exaggerated scratching of conspecifics might be interpreted as abnormal behaviour. Conspecifics might be used as substitute objects when litter is missing. Further investigations into this behaviour and possibilities to reduce agonistic behaviour especially in male rabbits are necessary.

Can additionally structured pens reduce agonistic interactions and injuries in male fattening rabbits?

Niebuhr, Knut, Frahm, Sandra, Waiblinger, Susanne, Smajlhodzic, Fehim and Windschnurer, Ines, Institute of Animal Husbandry and Animal Welfare, University of Veterinary Medicine Vienna, Department of Farm Animals and Veterinary Public Health, Veterinaerplatz 1, 1210 Wien, Austria; Knut.Niebuhr@vetmeduni.ac.at

Agonistic interactions and injuries can affect the welfare of male fattening rabbits. Partitions might reduce agonistic interactions by providing visual and physical separation in pens. This study compared the behaviour (e.g. agonistic, play, exploratory, sexual) and injuries of male rabbits group-housed either in minimally structured pens (only elevated platforms) or identical pens with elevated platforms and 4 additional wooden partitions (length × height: 3x 0.57×0.3 m; 1x 0.95×0.3 m) under/in front of the platforms. In total, 10 male rabbit groups (5 with/5 without partitions) were observed. 50 animals aged 5 weeks were placed in each pen (8 rabbits/m^2). The night before slaughtering (aged 82 days) the behaviour of each group was analysed continuously using video recordings (1 hour/group, at 21:00-22:00). The next morning 30 rabbits/group were examined for injuries due to agonistic behaviour (score 0 [no] to 3 [deep injuries]). The behaviour of groups with and without partitions (frequencies/group) was compared by means of T-Tests (pen as statistical unit, n=10). Injuries of individual rabbits were analysed by Chi^2-tests (n=300). The behaviour did not differ in pens with or without partitions (mean±SD/group×hour: Agonistic: without: 8.4±7.57; with: 11.6±12.86; P=0.644; Play: without: 0.2±0.45; with: 0.8±0.84; P=0.195; Exploratory: without: 46.2±15.55; with: 47.2±13.24; P=0.915; Sexual: without: 94.6±40.37; with: 58.8±22.19; P=0.120). There was also no difference regarding injured animals (without partitions: 61% without injuries; 24% deeper injuries; with partitions: 60% without injuries; 22% deeper injuries; Chi^2=0.759, P=0.859). Thus, no effect of partitions on agonistic behaviour or injuries was found. Maybe the additional partitions did not provide enough separation to reduce agonistic interactions. Groups in minimally structured pens showed more sexual behaviour and thus might be more disturbed, although a level of significance was not reached. This might have been due to the rather low sample size of n=5 per system. Further investigations, e.g. with additional structure modifications seem necessary.

Enrichment choice tests with laboratory mice: preferences during activity and resting

Heizmann, Veronika[1], Fritz, Martina[2] and Tichy, Alexander[3], [1]Institute of Animal Husbandry and Animal Welfare, Vetmeduni Vienna, Veterinärplatz 1, 1210 Vienna, Austria, [2]Praxisgemeinschaft am Geiselberg, Gottschalkgasse 17, 1110 Vienna, Austria, [3]AG Biostatistik, Institute of Population Genetics, University of Veterinary Medicine Vienna, Veterinärplatz 1, 1210 Vienna, Austria; veronika.heizmann@vetmeduni.ac.at

Improving the artificial environment of laboratory mice according to species-specific requirements and preferences should also improve the welfare of laboratory mice. A series of continuous choice tests was performed with Outbred mice to assess the value of different enrichment items in cages. 12 adult female Him:OF1 mice were tested in pairs, 12 adult (6 ♂/6 ♀) Him:OF1 mice were tested individually. The test-system consisted of 4 Makrolon-type3-cages (37×21×14 cm) connected by perspex tubes and a central corridor. Cages A,B,C,D were equally equipped with wood shavings, pellets in a dish and drinking water. In addition cages A,B,C were structured with nesting material (A,B,C), nestbox (A,C) and/or cardboard tubes (A,B). The mice were videorecorded and observed by instantaneous sampling on 3 consecutive days. At each sample point parameters from 3 categories were recorded: (1) behaviour; (2) location; (3) object of occupation. Relative dwelling times in cages and corridors were calculated as percentage of all observation points during the dark, light and dark-and-light periods. Depending on data distribution, measurements ANOVA and paired t-tests or nonparametric tests were applied. According to overall dwelling times the adult mice clearly preferred a cage with nesting material & nestbox: The 12 adult females spent most of the observation time in cage A, followed by cages C and B (cage effect: light: $P<0.01$ dark: $P<0.05$). The 12 adult ♂ and ♀ mice spent most of the observation time in cage C, followed by cages A and B (cage effect: light: $P<0.001$ dark: $P<0.01$ interaction cagexlight: $F_{3,30}=13.0$, $P<0.001$). At night the adult mice explored and used the whole testsystem including the unstructured cage D. The 12 male and female mice spent 11.5% (SD3.4) of the observation time running, 9.7% (SD1.8) eating, 4.0% (SD2.2) digging. Eating pellets from the floor made 4.5% (SD3.0). In addition to providing suitable nestsite&shelter, the artificial environment of laboratory mice should also enable species-specific locomotion, exploration and foraging.

Assessment of an environmental enrichment program for captive Southern pudu (*Pudu puda*)

Valdes, Beatriz, Caiozzi, Andrea and Zapata, Beatriz, Escuela de Medicina Veterinaria, Universidad Mayor, Unidad de Etología y Bienestar Animal, Camino La Piramide 5750, Santiago, Chile; andrea.caiozzi@umayor.cl

Pudu puda is a very small deer that inhabits forests in Chile and Argentina and is one of the less known South American deer. The behavior of the southern pudu (a shy, reclusive species) and the characteristics of its inhabited environment, hinder observation in the wild. Pudu are kept in captivity, however the complexity of their natural habitat is difficult to artificially recreate and their shyness turn them fearful and stressed. In this study we implemented a program of environmental enrichment (EE) in pudus to increase exploratory behaviors and the use of their space. Seven specimens were used in this study allocated in three enclosures. Ten hours of behavioral observations were conducted before (baseline) and 10 during the EE program, which was repeated twice. Feeding, exploring, resting, and social and stress related behaviors were recorded. Nutritional (N), sensorial (S) and occupational (O) items were given and measured separately. The room was divided in 'feeding zone', 'shelter zone' and 'neutral zone' to assess the use of the space. Repeated measures ANOVA with three levels were conducted to assess the effect of EE program. The three stimuli significantly increased exploration (eg.9.0±7.39 min vs. 18.1±11.33 min, after N) and decreased vigilance (eg.41.6±29.9 min vs. 14.4±23.1 min after N) ($P<0.05$). The response to N items was significantly higher than S and O items (12.35 min N vs. 8.7 min S and 2.6 min O) ($P<0.05$). Regarding the space use, EE program allowed animals to use the enclosures in a different way increasing the use of 'neutral zone' (26.1±14.99 min vs. 115.8±85.58 min) and decreasing the use of 'shelter zone' (565.2±22.09 min vs. 470.5±87.75 min) ($P<0.05$). We conclude that EE program applied in captive pudu had a positive impact promoting exploration and occupying the space in a more positive way.

Nest box color and height preference of White Leghorn laying hens raised in floor pens

Eusebio-Balcazar, Pamela, Didde, Dana and Purdum, Sheila, University of Nebraska-Lincoln, Animal Science, Lincoln, NE 68504, USA; pamelaeusebio@huskers.unl.edu

The objective of this study was to evaluate nest box color and height preference of laying hens. For this purpose, at 33 wk of age, 232 Bovan White laying hens were randomly allocated to eight wood shaving litter floor pens. Each floor pen contained 29 hens (0.167 m^2/hen) that were provided six plastic nest boxes (Chick Box TM, Broiler Equipment Company, Shropshire, England) arranged to provide two height levels (15 or 50 cm) and two colors (yellow or brown). Nest height was measured from the litter floor to the plastic perch attached to each nest box. Number of eggs laid in each nest box was recorded daily. Percentage of eggs relative to only nested eggs were analyzed using a completely randomized block design considering floor pens as blocks. More eggs were observed at the lower nest boxes (15 cm) than at the higher nest boxes (50 cm) (87.3% vs. 12.7%; SEM=0.97; $P<0.001$). Ten percent more eggs were laid in the brown nest boxes compared to yellow nest boxes (55.5% vs. 44.5%, SEM=1.83, $P<0.001$). Thus, White Leghorns laying hens moved from conventional cage systems to floor pen systems at 33 wk of age tended to prefer to lay in brown nest boxes at 15 cm compared to yellow nest boxes at 50 cm above the floor.

Outdoor use and forage intake of laying hens in mobile housing

Hörning, Bernhard, Trei, Gerriet, Schwichtenberg, Marcel, Kaiser, Tanja and Kallenbach, Elisa, University of Applied Sciences Eberswalde, Organic Animal Production, Friedrich-Ebert-Str. 28, 16225 Eberswalde, Germany; bhoerning@hnee.de

The aim was to investigate the use of the outdoor area by hens kept in mobile housing which is not a common system in Northwest Europe. 400 laying hens were kept in 2 mobile houses in 8 groups with 1 male per group. The houses were moved every 2 weeks (2.5 m^2 outdoor area per hen). Direct observations were carried out at 12 days in summer 2011 (each day 4 scan samplings in hourly intervals and 8 focal animal observations à 10 minutes each). Position of the animals in the outside area was recorded in 3 distance zones from the houses (4, 8, 12 m). Ground vegetation was analysed with the Braun-Blanquet method. Grass samples were taken at the beginning and the end of any moving of mobile houses and fresh weight and dry matter determined. Hypothetical grass intake per hen and day was calculated using the differences between these weighing. On average 28.0±8.8 hens were recorded in the outdoor area (47.6±23.4% in zone 1, 16.9±13.0% in zone 2 and 30.9±20.6% in zone 3). Main activity of hens in the run (scan sampling) was foraging, followed by walking, standing, preening, dustbathing and lying (70.3±18.0, 15.2±13.7, 6.1±11.1, 3.1±5.6, 2.6±6.2, 2.3±5.0% of animals). Focal observations of hens revealed a similar picture (e.g. foraging 66.7% of time). Males showed mainly standing, walking and foraging (32.1, 29.0, 23.8%, respectively). Vegetation heights were reduced clearly during the two week periods, mainly in the two lower herb layers, as well as cover ratio (from 77 to 38%), and number of plant species (from 10 to 8). Calculated grass intake per hen and day was 255 g fresh weight (57 g dry matter, 921 KJ ME). In conclusion, outdoor runs were used fairly well. Main activity of hens was foraging. Hens consumed a high amount of forage.

Influence of males on behaviour of laying hens in mobile housing

Trei, Gerriet, Hörning, Bernhard and Gnilke, Jurinde, University of Applied Sciences Eberswalde, Faculty of Landscape Management and Nature Conservation, Friedrich-Ebert-Straße 28, 16225 Eberswalde, Germany; Gerriet.Trei@hnee.de

The aim was to study possible influences of males on behaviour of hens in the outdoor run of mobile houses. 400 chickens were kept in 8 groups of 49 hens and 1 male in 2 mobile houses. After six months of lay, males were taken out of 4 groups for 1 month. Mobile houses were moved every 2 weeks (2.5 m^2 per hen outdoor area). Direct observations were carried out at 22 days in autumn 2011 (466 scan samplings of groups in 1 hour intervals and 118 focal observations of males for 10 minutes). Position of the animals in the outside run was recorded in 3 distance zones from the houses (4, 8, 12 m). Data were analysed with Mann-Whitney-U-test. A mean of 26.2±8.8 hens per group was in the outside run (scan sampling). Males were outside in 69% of observations. Main activities of hens were foraging (82.7±17.8%) and of males standing and foraging (54.1, 22.4%). Focal observations of males revealed similar results (58.2% standing, 27.6% foraging). There was no difference in hen numbers in the run between groups with or without males. However, more hens were in zone 1 nearest the house in group without males (53.7±20.9 vs. 49.9±25.3% of hens outside, P=0.026) and these hens showed less foraging (81.0±16.3 vs. 83.6±18.6%, P=0.006) and more standing (10.2±11.8 vs. 8.9±14.9%, P=0.001). In groups with males, more hens were outside if the male was present in the run (26.9±7.9 vs. 21.3±10.9, P=0.000). If no male was outside (56.1±28.7 vs. 47.5±23.5%, P=0.028), more hens were in zone 1 (56.1±28.7 vs. 47.5±23.5%, P=0.028) and less in zone 3 (24.1±20.9 vs. 29.9±20.8%, P=0.025). In conclusion, presence of males did not influence number of hens in the run. However, males influenced hen behaviour and distribution in the run.

Effect of group size and space allowance on selected behavioural and physiological measures of laying hens in enriched cages

Bilčík, Boris, Cviková, Martina and Košťál, Ľubor, Institute of Animal Biochemistry and Genetics, Slovak Academy of Sciences, Moyzesova 61, 90028 Ivanka pri Dunaji, Slovakia (Slovak Republic); Boris.Bilcik@savba.sk

Conventional cages are banned from 2012 being replaced mostly by enriched cages. Whether these cages are welfare friendly and what parameters they should meet is still subject of debate. In our study we focused on the effect of group size and space allowance on selected physiological and behavioural measures of laying hens in the DKG Hostivice enriched cage. We used 30 non-beak-trimmed ISA Brown laying hens, obtained as adult from the commercial breeder. Hens were kept in conventional cages until transfer to experimental facility. After, they were kept in the EU-125 enriched cages (DKG Hostivice, CZ) with the area of 7,500 cm^2. Two experimental group sizes/space allowances were used: 10 birds (recommended group size, 750 cm^2/ hen) and 5 hens (1,500 cm^2/hen). Hens were fed *ad libitum*. Behavioural observations (direct observations with scan sampling and video recording) focused on the use of space and resources, resting, aggressive and feather pecking, and exploration. Physiological recordings were done using the radiotelemetric system (Data Sciences International Inc., USA), with surgically implanted transmitter TL11M2-C50-PXT. System allowed us to measure motor activity, ECG, body temperature, heart rate and blood pressure in freely moving animals. Birds from smaller group size (5 hens) displayed higher frequency of aggressive pecking as well as severe feather pecking and were more active. Birds from the larger group size (10 hens) spent more time eating and standing. Use of the nest box differed as well – 88% of eggs in smaller group were laid in the nestbox, compared to only 45% in larger groups. There was a trend towards lower blood pressure and body temperature in smaller group size birds. Contrary to expected results, some of the measured parameters at lower stocking density seemed to be worse than in hens kept at higher density. Currently we are collecting more physiological and behavioural data to validate our results. This work was supported by the APVV-0047-10.

How do laying hens use the offered perches in modern German colony cages?

Brügesch, Femke, Fels, Michaela, Spindler, Birgit, Lohan, Katja and Hartung, Jörg, Institut for Animal Hygiene, Animal Welfare and Farm Animal Behaviour, University of Veterinary Medicine Hanover, Foundation Germany, Bünteweg 17p, 30559 Hannover, Germany; femke.bruegesch@tiho-hannover.de

European legislation (1999/74/EU) requires the provision of nestboxes, dustbathing areas and perches in housing systems for laying hens. This study is part of a larger research project investigating the animal welfare benefits of new German Colony Cages (890 cm^2/hen). The presented experiment should clarify whether or how laying hens use the offered perches in the new housing system. Therefore, the use of perches by layers was video-recorded for two days in the beginning, in the middle and at the end of a laying period, respectively. A total of three cages (each with 36 LSL hens), representative for all cages in the stable were filmed. The use was registered once per hour (14 h) during light phase, twice by night and expressed as % of total observations. ANOVA analysis followed by posthoc test was conducted for total perch use. Wilcoxon-test was performed for testing preferences for different perches. Our results show that the use of perches during the light phase varied in the course of laying period. In the beginning, the use of perches was lower than in the middle and at the end of the laying period (6% vs. 14% and 17%, $P<0.05$), whereas the use at night was higher in the middle than in the beginning and at the end of the laying period (71% vs. 49% and 57%). The most preferred perch during daytime was the lower plastic perch followed by the higher steel perch and the steel dust refilling tube (50% vs. 28% and 22%, $P<0.05$). The latter was primarily used to reach the nipple drinkers. At night, the majority of birds roosted on lower plastic perch (47%) and higher steel perch (46%). The use of perches at night was lower than expected. Nevertheless, these first results demonstrate that perches can contribute to the enrichment of the hens' environment.

Does dustbathing behavior by laying hens reduce Northern Fowl mite populations?

Vezzoli, Giuseppe and Mench, Joy, University of California, Davis, CA, USA, Department of Animal Science, One Shields Avenue, 1403 Mench Lab, 95616, USA; gvezzoli@ucdavis.edu

Northern Fowl mites are common ectoparasites of laying hens. Infestation causes skin irritation, and severely infected hens can become anemic. It has been suggested that dustbathing helps to remove ectoparasites. However, there are no published studies evaluating this presumed function of dustbathing using natural, untreated substrates. We investigated whether providing hens with a natural substrate that is highly preferred for dustbathing was effective in reducing Northern Fowl mite (*Ornithonyssus sylviarum*) populations. Beak-trimmed White Leghorn hens (n=16) were singly housed in 0.4 m^2 cages provided with 0.15 m^2 plastic dustboxes that were either empty (CONTROL) or filled daily with 1,200 g of sand (SAND). At 25 weeks of age each hen was infested with 35 mites. The hens were scored for mite populations visually using a 0-7 scale weekly, and video-recorded for 8 hours for 2 consecutive days immediately before infestation and 1, 3, 5 and 7 weeks after infestation. Mite population data were analyzed using the Wilcoxon test and dustbathing behavior using the Kruskal-Wallis test. There were no significant differences in mite populations between treatments at any time point post-infestation (all Z values were ≥0.64 and p values ≥0.20). The peak of infestation occurred at 5 weeks in both CONTROL (median score = 5.5±2.7 CI) and SAND (median = 6.0±1.7); scores of 5 and 6 correspond to 501-1000 and 1,001-10,000 mites on the ventral abdomen. At the peak of infestation the total duration (median = 617 sec ± 1,940 CI) and number of bouts (median = 1bout ± 1.73 CI) of dustbathing behavior did not change (H_2=1.14, P=0.56; H_2=1.44, P=0.48, respectively) in SAND hens compared to the pre-infestation baseline. These data provide new evidence that dustbathing is not elicited by the mite load on the hen's integument and does not help to remove mites, at least when the substrate provided is sand.

Exploratory and dust-bathing behaviour in laying hens kept in aviary systems – litter management matters

Döring, Susanne, Staack, Marion, Brenninkmeyer, Christine and Knierim, Ute, University of Kassel, Farm Animal Behaviour and Husbandry, Nordbahnhofstr. 1a, 37213 Witzenhausen, Germany; doering@uni-kassel.de

The possibility to perform exploratory and dust-bathing behaviour substantially contributes to an improved welfare state of laying hens. This study including 10 commercial aviary houses aimed to determine possible effects of litter height, type and humidity, flock size, stocking density and light intensity on the quantitative expression of these behaviours. Three to four cameras/house each covered 1 m^2 of the indoor scratching areas. Instantaneuous scan sampling every 2 minutes over the first quarter of each lighting hour, distributed over 2 days was carried out, light intensity and litter height measured and litter quality scored. Litter height ranged from 0 to 8 cm, light intensity from 4 to 14 Lux. Litter type was assessed as either missing, straw, straw mixed with fine material or fine material. Flock sizes per pen ranged from 400 to 5,000 birds, stocking densities from 6.1 to 18.5 hens per m^2. On average each of the 6.5±4.2 animals within the observed areas showed pecking during 25.3%±10.3, scratching during 2.6%±1.6 and dust-bathing during 7.0%±7.7 of observation time. More hens were observed in the scratching areas in larger flocks (proc mixed, df=30, P=0.049), with higher stocking density (P=0.000) and a tendency was seen depending on litter type (P=0.068), namely on straw versus missing or fine material. There was a tendency that hens scratched more on higher litter (P=0.052). No effects on pecking could be identified. Dust-bathing was performed more often when litter was dry (P=0.031) and there was a tendency with greater litter height (P=0.063). Also litter type had an effect (P=0.018), with more dust-bathing on straw than on mixed or missing litter. Results indicate that within the range of commercial conditions found, especially litter management affects the extent of exploratory and dust-bathing behaviour. The work will be continued.

Possible influences on feather pecking in laying hens housed in two types of aviary systems

Smajlhodzic, Fehim, Arhant, Christine, Wimmer, Andreas, Troxler, Josef, Zaludik, Kathrina and Niebuhr, Knut, Institute of Animal Husbandry and Animal Welfare, University of Veterinary Medicine Vienna, Department for Farm Animals and Veterinary Public Health, Veterinärplatz 1, 1210 Vienna, Austria; Fehim.Smajlhodzic@vetmeduni.ac.at

Feather pecking can be a serious welfare problem in laying hens. Studies regarding feather pecking of hens housed in large aviaries are rare. Therefore, possible influencing factors on feather pecking (assessed by feather damage) in non beak-trimmed hens were investigated in two types of aviary systems (portal: n=23/row: n=25, 48 flocks visited in total). At the age of 37±3 (mean±S.D.) weeks (in the laying peak), on each farm 30 hens (in total 1,440 hens) were scored (examining neck, back, vent, wings, pinion and tail feathers) and for each farm an average feather damage score was calculated, ranging from 0 (no damage) to 19 (highest possible score). Factors examined were: type of aviary system (portal or row), feeding management (frequency), litter (dry and flaky, partly sticky, compacted), space allowance after placement and difference of actual weight from desired weight, experience of the farmer in years, pecking behaviour towards humans (pecking against plastic overshoes within first 10 minutes after entering the house, yes/no). The feather damage score ranged from 1.03 to 8.47 (mean±S.D.: 3.39±1.66). Stepwise backward regression was performed to identify which variables were the best predictors for plumage damage. The final model explained 19.3% (R^2) of the variance (P=0.008). Remaining predictors were: number of years the farmer had run their aviaries (β-coefficient: 0.243, P=0.076) and feeding frequency per day (β-coefficient: -0.364, P=0.009). Other influencing factors, such as the type of system, had no effect. The higher the feeding frequency, i.e. the more often per day hens were fed, the lower was the average feather score. Yet, the longer the farm owners had run their aviaries, the higher was the feather pecking score. This may be due to the fact that after some years farmers pay less attention to management details or due to less favourable characteristics of older aviary systems.

How does feather eating affect intestinal microbiota and microbial metabolites in growing leghorn-type chickens?

Meyer, Beatrice[1], Vahjen, Wilfried[1], Zentek, Jürgen[1] and Harlander-Matauschek, Alexandra[2], [1]Free University of Berlin, Institute of Animal Nutrition, Brümmestrasse 34, 14195, Germany, [2]Vetsuisse Faculty Bern, VPH Institut, Division Animal Welfare, Länggassstrasse 120, 3012 Bern, Switzerland; beatrice.meyer@fu-berlin.de

In laying hens it is known that feather pecking is associated with feather eating. Feathers are almost resistant to gastric and pancreatic proteolytic enzymatic degradation. However, feathers might serve as a substrate for the gut microbiota and influence bacterial colonization. It is reported that nutrition modulates gut microbiota and that events occurring in the gut also have an impact on behaviour. Regarded as a first step we determined the influence of feathers in the diets of chicks on the composition of the intestinal microbiota and their metabolites. 61 one-day-old Lohmann-Selected Leghorn chicks were divided into 3 diet treatment groups: Group A (control) and B (5% chopped feathers in the diet) included 24 birds each and C (13 birds), in which the control diet was fed until week 12 and then switched to diet B. From group A and B 12 birds each were slaughtered in wk 10 and wk 17. Group C was slaughtered in wk 18. Digesta was taken from ileum and caecum and microbiota identified by cultivation and DNA sequencing. Microbial metabolites (short-chain fatty acids, ammonia and lactate concentrations) were recorded. There was a significantly increased growth of keratinolytic bacteria in groups B and C in the ileum ($P<0.001$) and caeca ($P=0.033$), which indicates that the time of exposure to chopped feathers did not influence the adaptation of keratinolytic bacteria. Keratinolytic bacteria were identified as Enterococcus faecium, Lactobacillus crispatus, Lactobacillus reuteri-like species and Lactobacillus salivarius-like species. The bacterial metabolism was slightly affected by feather chopped diets in the caeca, where higher ammonia and short-chain fatty acids but lower L-lactate concentrations were found. This experiment showed that ingested feathers affect intestinal microbiota. Our findings may stimulate research into the interaction between feather eating, the adaptation of intestinal microbiota and the impact on feather pecking in layers.

Measurement of floor space covered by pullets in various body positions and activities

Spindler, Birgit[1], Briese, Andreas[2] and Hartung, Jörg[1], [1]Institute for Animal Hygiene, Animal Welfare and Farm Animal Behaviour, University of Veterinary Medicine Hanover, Foundation, Germany, Bünteweg 17p, 30559 Hannover, Germany, [2]Bri-C-Veterinärinstitut, Gartenstraße, 31157 Sarstedt, Germany; birgit.spindler@tiho-hannover.de

The available floor space has a strong impact on the execution of various behaviours of laying hens. Presently there are no specific legal requirements on the keeping of pullets (0 to 18 weeks of age) and no minimum floor space is defined which may be sufficient for their behavioural needs. In order to get a first approximation, the floor space covered by pullets in various body positions and during some activities was determined by the colour contrast planimetric method 'KobaPlan'. In total 1,550 hens from two brown feathered (Lohmann Brown -LB- and Lohmann Tradition -LT-) and two white feathered hybrids (Lohmann Selected Leghorn -LSL- and Dekalb White) were measured from the 6th week of life to 18 weeks at regular intervals. The hens were weighed and photographed digitally in a box and in a sand bath that were lightened and floored to provide a high contrast background. The KobaPlan program counts the number of pixels of each picture and calculates the animal area. The mean floor space covered by LB and LT at the end of rearing (body weight 1,450±25 g) was 422±42 cm^2 and 432±70 cm^2 standing, respectively 448±51 cm^2 and 445±43 cm^2 sitting. LSL and Dekalb White (body weight 1,300±25 g) used 371±41 cm^2 and 349±26 cm^2, respectively in standing and 379±41 cm^2 sitting. Wing flapping with up to 1,755±504 cm^2 was the most space-occupying behaviour, followed by dustbathing (737±80 cm^2) and preening (690±49 cm^2). The planimetric survey gives a first allometric basis for the estimation or even calculation of the area needed by pullets when sitting, stretching and other behaviours. Further experiments should be related to specific ethological observations.

Aspects of broiler chicken welfare on free range farms in Southern Brazil

Sans, Elaine C. de O.[1], Federici, Juliana F.[1], Dahlke, Fabiano[2] and Molento, Carla F.M.[1], [1]Universidade Federal do Paraná, Animal Science, Rua dos Funcionários 1540, 80035-050, Brazil, [2]Universidade Federal de Santa Catarina, Animal Science, Campus Reitor João David Ferreira Lima Trindade, 88040-970, Brazil; carlamolento@yahoo.com

Our objective was to assess broiler chicken welfare in free range units in the State of Paraná, Brazil. Ten units were visited at most five days before slaughter, and for five of them slaughter was monitored. Welfare was assessed according to the Welfare Quality® for Poultry. Results are presented as median (min-max) ranging from 0 to 100 (the higher, the better the welfare) except for feelings. Index for number of drinkers was 93 (41-100); plumage cleanliness 100 (95-100) or 98.8% clean; litter quality 34 (14-67) or leaves imprint of foot; dust 53 (20-53) or thin dust covering; thermal comfort 100 (100-100) or 0% panting/huddling; stocking density 56 (26-88) or from 9.3 to 38.7 broilers/m^2; lameness 81 (63-98) or 35.5%; hock burn 93 (83-99) or 3.6%; footpad dermatitis 35 (8-70) or 54.0%; mortality 2.0 (1.4-7.2%); avoidance distance test from broiler to human 70 (25-100); medians for positive feelings ranged from 48 to 93, and for negative feelings from 20 to 80. Results for slaughterhouse measurements were mean duration of total (farm, transport, and waiting area) water withdrawal 213±38 min, and of total feed withdrawal 861±85 min; emaciated broilers 97 (74-100) or 0.08%. No wing damage, dehydration or ascites was observed. There was 15.3% (7.0-17.9%) wing flapping on the line; 9.3% (6.7-16.7%) bruising; 4.5% (1.8-11.0%) hepatitis; 0.08% (0.00-0.10%) abscesses; 49.3% (26.7-56.8%) pre-stun shock; 3.9% (1.3-6.7%) ineffectively stunned broilers. Pre-stun shock was immediately looked after. Results suggest that there are important opportunities for welfare improvement, and that low lameness and good thermal comfort present a favorable scenario for free range broiler chickens welfare.

Contact behaviour of shelter dogs towards strangers – relationships with characteristics of the dog and the shelter environment

Arhant, Christine, Kadlik, Sandra and Troxler, Josef, Institute of Animal Husbandry and Animal Welfare, Department for Farm Animals and Veterinary Public Health, Veterinärplatz 1, 1210 Wien, Austria; Christine.Arhant@vetmeduni.ac.at

Behaviour of shelter dogs towards strangers might influence adoption rates. Therefore, we investigated factors related to positive contact behaviour regarding individual characteristics of dogs or shelters. In 29 shelters behaviour of dogs towards strangers was tested by an approach test in front of the kennel. Successful 'Contact' was assigned if a dog approached and explored the experimenter. 'NoContact' was assigned if the dog ignored, attacked or ceaselessly barked or growled at the experimenter (inter-rater-reliability: Kappa=0.86, $P<0.001$, n=158). Influences of dog characteristics (age, size, sex, duration of stay) and the presence or absence of shelter-staff during the test were analysed on an individual level (n=552). Relationships between shelter environment and dog behaviour were analysed on shelter level calculating the percentage of dogs showing contact behaviour/shelter (test-retest-reliability [9 shelters visited twice]: $r_s=0.76$, P=0.017). Kennel environment was assessed using score sheets. Routines of dog care and staff human-animal-relationship were surveyed by questionnaires. A median percentage of 79% dogs/shelter showed contact behaviour (Min: 33%, Max: 100%). On individual level only the duration of stay in the shelter was related to contact behaviour (Contact: Median 6 Month, NoContact: Median 10 Month, Z=-3.61, $P<0.001$). The presence of staff during the test did not influence dog behaviour (P=0.217). In larger shelters less dogs showed contact behaviour ($r_s=-0.62$, P=0.001). A higher percentage of dogs showing contact behaviour was related to a higher percentage of kennels with soft bedding ($r_s=0.48$, P=0,012), providing toys regularly (Z=-2.10, P=0.036), feeding more fresh meat ($r_s=0.45$, P=0.016) and cleaning kennels more often ($r_s=0.44$, P=0.030). Additionally, a higher agreement of staff to be friendly, patient and predictable when handling dogs was related to increased contact behaviour ($r_s=0.45$, P=0.047). These findings suggest that good care and a positive human-dog-relationship might lead to dogs being more willing to contact strangers and thus increase adoption rates.

Dogs' behavioural responses in human-dog interactions – influences of being familiar or unfamiliar to the pet

Kuhne, Franziska[1], Hößler, Johanna C.[2] and Struwe, Rainer[2], [1]Animal Welfare and Ethology, Department of Veterinary Clinical Sciences, Justus-Liebig-University Giessen, Frankfurter Strasse 104, 35392 Giessen, Germany, [2]Institute of Animal Welfare and Behaviour, Veterinary Department, Freie Universität Berlin, Oertzenweg 19b, 14163 Berlin, Germany; franziska.kuhne@vetmed.uni-giessen.de

Understanding the behavioural responses of individual dog in human-dog interaction is crucial for interpreting the risk posed by a given dog. Dogs usually differentiate in social communication between pack and non-pack members. The physical part of dogs' communication is used to maintain social affiliation or to impress an opponent. Therefore, we examined if the human-dog familiarity has far-reaching influences on dogs' behavioural responses. The participating dogs (n=24) were privately owned pet dogs of different breeds and either gender, aged 1 to 11 years. Each dog was exposed to nine different human-dog interactions for 30 seconds in a normal office setting (e.g. petting the dog's paw or head). The inter-test interval was set at 60 seconds. The person was always the same woman who was instructed to give no other, potentially threatening signals. The person was either familiar or unfamiliar to the dog. The dogs of the familiar group could previously recognise the person while frequently walking by their owners within a human-dog group. The frequency and duration of the dogs' behavioural responses on each interaction were analysed. The analysis of variance was conducted after log data transformation of the behavioural responses was carried out. A significant influence of the human-dog familiarity on behavioural responses was found for initiating redirected behaviours (e.g. sniffing/licking on the floor) ($F_{1,16}$=4.9; P=0.027) and for appeasement gestures (e.g. averted gaze or head), both in frequency ($F_{1,19}$=10.7; P=0.001) and duration ($F_{1,18}$=21.8; P=0.000) in that dogs being petted by a familiar person intensified these behavioural responses. Our results show that the human-dog familiarity has an effect on dogs' appeasement gestures and redirected behaviours to tactile human-dog interactions. Additional study is needed to assess the human handler awareness of these behaviour patterns and determine whether the dogs' responses are potential risk indicators of human-dog interactions.

Dogs hunting bears – what do they really hunt?

Forkman, Björn[1], Gustavsson, Tobias[2], Støen, Ole-Gunnar[3], Karlson, Jens[2] and Brunberg, Sven[4], [1]University of Copenhagen, Dept Large Animal Sciences, Grønnegårdsvej 8, 1870 Frederiksberg C, Denmark, [2]Swedish University of Agricultural Sciences, Grimsö Research Station, 730 91 Riddarhyttan, Sweden, [3]Norwegian University of Life Sciences, Dept Ecol Nat Res Managm, P.O. Box 5003, 1432 Ås, Norway, [4]Scandinavian Bear Project, Tackåsen Kareliusvag 2, 794 98 Orsa, Sweden; bjf@life.ku.dk

Animals wounded in accidents or in hunts are a serious welfare problem, and should be found and killed as soon as possible. The hunting of the animal starts with a dog being shown the track, the dog tracking the prey while on leash, it is let go and is expected to pursue the prey and finally to start baying. In the current study 22 experienced hunting dogs were used. Fourteen of the dogs had previously been used for hunting bears. Eleven of the dogs had participated in a bear hunting test (developed and used by the Swedish hunting association). The evaluation includes accessibility, reaction when in contact with the bear (through a fence), independence, bark, sharpness, and recall. In the actual hunt wild bears with gps collars were used as prey animals. All hunting dogs had gps collars as well; this enabled us to find a recent bear track as well as gathering data on how closely the dog was following the bear. Out of the 22 dogs, seven did not start tracking or showed a very weak interest in the track. Only two dogs found the bear and kept it at bay until the dog could be recovered by the hunter, only these individuals can be said to have succeeded in the hunt. When scoring the success in the hunt the dogs having succeeded in the bear test did not show a better result than those had failed them ($P=0.46$, $n1=4$, $n2=7$). The results indicate that many dogs are poor at hunting bears, especially when they are required to follow a specific individual and that other training and testing methods than the ones currently employed should be used.

Owner-support influences conflict behaviour differently in friendly and aggressive dogs

Beerda, Bonne[1,2], Neessen, Petra[2], Boks, Sabine[2], Alderliesten, Evelien[2], Van Staaveren, Nienke[2], Li, Xun[2], Van Zeeland, Renate[2], Van Adrichem, Anita[2] and Van Der Borg, Joanne A.M.[1], [1]Wageningen University, Department of Animal Sciences, Adaptation Physiology Group, P.O. Box 338, 6700 AH Wageningen, the Netherlands, [2]Wageningen University, Department of Animal Sciences, Behavioural Ecology Group, P.O. Box 338, 6700 AH Wageningen, the Netherlands; bonne.beerda@wur.nl

Undue fear-aggression in dogs compromises welfare and affects society via biting-incidences. The Netherlands accommodates ~1.5 million dogs and 16.7 million people, and faces annual incidences of 8 dog-bites per 1000 citizens. Animals respond emotionally to human presence, and dog-owners possibly influence their dogs' behavioural expression of fear-aggression (including biting). We tested the latter by observing (focal continuous) owner-dog pairs (n=66) that were approached by an unfamiliar odd looking and acting person (equipped with mask and walking-stick). Approach, stand-still (5 s, 2 m distanced) and retreat lasted about 25s and Trials were repeated thrice with minute intervals. Owners were instructed to Actively Support (AS; speaking calmly and petting from head-to-tail) the dog or Stand Passively (SP). They reported on their dog's home behaviour by answering questions on five-point scales, generating scores for Stranger-Directed Aggression (SDA). Principal component analyses revealed 3 dog behaviour dimensions, each explaining % variations over 10 and with relevant loadings >|0.45|, namely: Arousal-Social (grouping the behaviours interact_with_owner, look_at_owner, tail-wag, tongue-flick and, reversely, tail-low), Aggression (bark, growl, piloerection_or_lunge, high-tail, look_at_stranger) and Approach (approach_stranger, move, pull_leash). Component scores (198 records) were analysed using Linear Mixed Models (random effect=dog) with the fixed effects Trial[1,2,3], SDA[covariate expressed as % of max] and Support[AS,SP], including 2-way interactions. Active support caused high scores for Arousal-Social (0.3±0.2 vs. SP -0.8±0.2, P<0.001) independent of SDA scores, and low scores for aggression in friendly (SDA=0%) dogs (-1.0±0.3 vs. SP 0.1±0.2, Support×SDA P=0.034) with opposite effects in dogs with high (80%) SDA scores (3.1±1.2 vs. SP 0.8±0.7), possibly by disinhibiting behaviour. Dogs with high SDA scores tended to increasingly approach the stranger with repeated Trials, which opposed responses in non-aggressive dogs (Trial×SDA P=0.051). Owner-behaviour directly influences dogs' responses to 'threats', with owner-support triggering different behaviours in dogs that are friendly and aggressive and having a risk of releasing overt responses in aggressive dogs.

An observation study of the Campus Canines Program at the University of Pittsburgh

Camaioni, Nicole, University of Pittsburgh, School of Education, 5500 Wesley W. Posvar Hall 230 South Bouquet Street Pittsburgh, PA 15260, USA; nic10@pitt.edu

The Campus Canines Program (CCP) at the University of Pittsburgh offers a unique animal-assisted activity (AAA) experience by providing the University of Pittsburgh's community with the opportunity to interact with registered Canine Good Citizen dogs. My objective was to observe and describe the CCP to try to determine what happens during an AAA program at a major university. A single case (CCP) was explored. Data was collected through participant observation. Because this is the first attempt to study the CCP, this immersion observation was necessary to gain a baseline description of the social activities of the program. During five weeks, 45 participants, 15 volunteers, and 10 canines were observed. The observations were conducted in the Cathedral of Learning at the University of Pittsburgh for one hour each week. The results identified a recurring interaction pattern between the participants, volunteers, and canines. This interaction pattern begins by the canine eliciting participation by soliciting the participant to sit, pet the dog, and join the dog/volunteer team and other participants. Once the participant joins the group, the human-to-human connections occur through shared participant pet stories and shared volunteer pet stories. The mean duration of an instance of this interaction is nine minutes. Also notable in the results was the observation of the creation and maintenance of relationships. The four main types of relationships observed were participant – participant, participant-volunteer, participant-family, and volunteer-volunteer. These connections were made through the sharing of stories and pictures of the their personal pet experiences. The observed interaction pattern provides a description of the social activities of the CCP. Also observed was the creation and maintenance of relationships between participants, volunteers, and family members. Therefore, this program may help to serve a need in providing social support for students in higher education.

Impact of rabies control strategy on people's perception of roaming dogs

Sankey, Carole[1], Häsler, Barbara[2] and Eckman, Harry[1], [1]World Society for the Protection of Animals, Programmes department, 5th floor, 222 Grays Inn road, London WC1X 8HB, United Kingdom, [2]Royal Veterinary College, University of London, Hawkshead lane, North Mymms, Hatfield, Hertfordshire, AL9 7TA, United Kingdom; carolesankey@wspa-international.org

The human-dog relationship has many faces. In most of Asia, a dog's ownership status is often unclear, as privately or community-owned dogs roam alongside unowned animals. Dogs can be a great enrichment in people's lives but they can also cause severe nuisance (e.g. bites, breeding, fouling), especially in parts of the world where rabies is endemic. Inhumane dog culls are often carried out in a misguided attempt to regulate roaming dog populations and control rabies outbreaks. The aim of this study was to investigate the impact of the implementation of a humane rabies and dog population control programme, including mass vaccination and targeted sterilization of dogs and public education, on the population's acceptance of dogs in Colombo City, Sri Lanka. Changes in people's acceptance were assessed using data collected in surveys in 2007 (1,823 households) and 2010 (975 households), as well as data collected in focus groups (61 participants in 9 focus groups) in Colombo City. Results show a higher acceptance of dogs in dog-owners than non dog-owners both in 2007 (Mann-Whitney U-test, $P<0.001$) and 2010 (Mann-Whitney U-test, $P<0.001$). However, the implementation of the rabies control programme led to a significant increase in dog acceptance in non dog-owners from 2007 to 2010 (Wilcoxon, $P<0.01$). In addition, all focus groups perceived a decrease in the number of problems caused by dogs, as well as in the actual number of roaming dogs, which contrasts with the stable numbers recorded by teams of the Blue Paw Trust (local animal welfare organisation). In conclusion, even though the roaming dog numbers remained stable throughout the intervention, dogs were perceived to be fewer and less of a problem, which could be explained by a possible improvement of their health status and better behaviour (e.g. less biting, breeding…), thus reflecting the success of the on-going control programme.

Trailer-loading in horses: influences of human behaviour and ramp material

Wiegand, Katharina and Gerken, Martina, University Goettingen, Department of Animal Sciences, Albrecht Thaer Weg 3, 37075 Goettingen, Germany; mgerken@gwdg.de

Difficulty of loading is a common problem in horses. Possible determinants of loading failure were studied to mitigate the problem. The first trial investigated the influence of human behaviour. Eight female persons led four horses in a fixed course into the same horse trailer. The loading of the same horse by the same person was repeated four times in a latin square design (total 128 loadings, four loadings per person/day). Trials were filmed and analysed for human behaviour such as orientation of eyes, lead tension, body position towards the horse, body posture, distance to horse, and use of voice. There was a significant influence of the individual human on the loading behaviour of the horse (Friedman test considering repeated measurements, $P=0.045$). The highest relationships (Cramér's V, $P<0.001$) between human behaviour and horse behaviour were found for orientation of eyes ($V=0.52$), lead tension (0.36), and body posture (0.36). To assess the effects of different floor materials of the loading ramp, the usual ramp floor (black rubber) was compared with straw and plastic turf. Eight horses were led by the same person three times across the three ramp designs (total: 72 loadings, three loadings per horse/day). On the first day, latency until stepping unto the ramp was higher ($P<0.05$) for the plastic turf. Across three repetitions, however, ramp material had no significant influence on loading time. Horses quickly habituated to repeated loading and novel materials. However, results should be replicated with different populations of horses and humans. Regular loading training might help to overcome loading problems. In the present trials, however, horses were not exposed to subsequent transport. The separate training of loading only might help to disrupt the conditioned avoidance learning, when the trailer has become associated with unpleasant experiences during travelling.

The effect of gentle handling treatment on avoidance distance and meat quality in extensively reared beef cattle

Hillmann, Edna[1], Probst, Johanna K.[1,2], Leiber, Florian[1], Kreuzer, Michael[1] and Spengler Neff, Anet[2], [1]ETH Zurich, Institute of Agricultural Sciences, Universitätsstrasse 2, 8092 Zurich, Switzerland, [2]FiBL, Research Institute of Organic Agriculture, Animal Husbandry Division, Ackerstrasse/ Postfach, 5070 Frick, Switzerland; edna-hillmann@ethz.ch

This study investigated the effect of two handling activities (based on TTouch®) in extensively reared and, therefore, possibly skittish beef cattle on stress related parameters on farm and at the abattoir. Ten Limousin crossbred beef cattle aged 15 months and from one herd were assigned to a handling group (HG=1 female, 4 steers) and a control group (CG=2 female, 3 steers) by pairing animals with similar avoidance distances. Back-region handling, comprising stroking and brushing the region of M. gluteus medius and biceps femoris and tail head, were carried out 10 times for 8 min during 5 weeks, starting 12 weeks before slaughter while animals were fixed in the feed barrier. Subsequently, head-neck-region handling was conducted 10 times for 8 min during 5 weeks, starting 5 weeks before slaughter. Before handling started and after all handling treatments, each animal went through an avoidance distance test (ADT). Blood samples collected during exsanguination were analysed for serum cortisol, lactate, and glucose concentrations. After slaughter, samples of M. longissimus dorsi were stored for 3 weeks and then tested for meat quality (cooking loss, shear force and colour). Data were analysed by using the two-sided Mann-Whitney-U-test (exact). Differences between first and second ADT tended to be greater in HG animals (W=67, P=0.056). Avoidance distance got much lower in HG animals (mean=70 cm, SE=58.31) than in CG (mean=160 cm, SE=62.29) animals. The handlings had no influence on exsanguination blood parameters. Shear force tended to be lower in meat from HG animals (W=4, $P<0.1$; mean=28.71 N, SE=0.29) than in CG animals (mean=30.66 N, SE=0.79). A gentle handling, applied on extensively reared animals, starting 12 weeks before slaughter, resulted in reduced avoidance distance and tended to increased meat tenderness.

The effect of training and milking familiarity on heifer behavioural reactivity to humans

Sutherland, Mhairi A. and Huddart, Frances J., AgResearch Ltd, Ruakura Research Centre, Hamilton, 3240, New Zealand; mhairi.sutherland@agresearch.co.nz

The objective of this study was to determine the effect of training and milking familiarity on behavioural reactivity of heifers towards humans. Behavioural reactivity to humans was assessed pre-partum (Test 1) using four behavioural tests: restraint, exit speed, human avoidance distance (HAD) and a human voluntary approach (HVA) tests. One month before calving, heifers were either trained in a rotary milking-parlour (n=20) or left undisturbed in the paddock (control, n=20). Training involved introducing heifers to the milking platform, associated noises and human contact (e.g. gentle handling/rubbing of the back of the legs and udder) during four sessions conducted over two consecutive days. The four reactivity tests were repeated immediately after (Test 2) and 12 wk after training (Test 3). Data were analyzed using the MIXED procedures of SAS. The repeatability estimates for response to restraint, exit speed, HAD and HVA tests between Tests 1, 2 and 3 were 0.23, 0.52, 0.13 and 0.14, respectively. The number of steps heifers took while restrained was similar (P>0.05) between Tests 1 and 2, but lower (P<0.001) in Test 3 [25.6, 25.4, 5.6±2.7 steps, Tests 1, 2, and 3, respectively]. The time taken for heifers to exit the restraint was similar (P>0.05) in Tests 1 and 2, but longer (P<0.001) in Test 3 [2.3, 2.3, 4.8±0.32 s, Tests 1, 2, and 3, respectively]. The avoidance distance of heifers to an approaching human was shorter (P<0.05) in Test 2 and 3 than Test 1 [4.9, 4.1, 2.9±0.2 meters, Tests 1, 2, and 3, respectively]. The distance that heifers voluntarily approached a stationary human did not differ (P>0.05) between test periods. Familiarity with the milking process but not training reduced the behavioural reactivity of heifers to humans, suggesting that the training regime used in this study was not adequate to reduce the behavioural reactivity of heifers.

Is placentophagia in goats correlated with good mother-young bonding?

Nandayapa, Edith[1], García Y. González, Ethel[1], Álvarez, Lorenzo[2], Flores, Alfredo[3], Ayala, Karen[1] and Terrazas, Angélica[1], [1]F.E.S.C Universidad Nacional Autónoma de México, Ciencias Pecuarias, km 2.5 Cuautitlán – Teoloyucan, San Sebastian Xhala, Cuautitlán Izcalli, Edo. Méx., 54714, Mexico, [2]F.M.V.Z. Universidad Nacional Autónoma de México, Centro de Enseñanza, Investigación y Extensión en Producción Animal en Altiplano, Km. 8.5 Carretera Federal Tequisquiapan a Ezequiel Montes.Querétaro, 76790, Mexico, [3]Universidad Autónoma Agraria Antonio Narro, Centro de Investigación en Reproducción Caprina, Periférico Raúl López Sánchez y Carretera a Santa Fe, Torreón Coahuila, 27054, Mexico; end370@hotmail.com

Placentophagia refers to the consumption of the placenta after giving birth. It is practiced by many mammalian species and its purpose is not clear. The objective of this study was to explore and describe the placentophagia behavior in post-parturient goat and their possible role in the ethogram of maternal behavior in this species. A total of 12 multiparous dairy goats (that gave birth to 4 singles, 6 twins and 2 triplets) were observed during the first 3 hours after birth. The behavior of the kids and the mothers were recorded and then analyzed using The Observer Video Pro (XT), Pearson chi-squared test was performed to compare proportions. We found that 9 of the 12 goats ingested the placenta; and no health problem was observed in those goats. The proportion of goats that ingest placenta tended to be higher (P=0.08) than those that did not. The latency to ingest the placenta was 86.2±10.85 minutes. The frequency of this behavior was 5.6±1.9 in a period of 30 minutes. The time spent by the dams ingesting the placenta was 605.8±214 seconds. A positive correlation between the latency to ingest the placenta and the latency of licking the second kid (r=0.7, P=0.03) was observed. A positive correlation was found between the frequency of ingest the placenta and the frequency of attempt to raise of the second kid (r=0.77, P=0.02). A tendency (P=0.1) for a positive correlation was found between the frequency of consuming the placenta and the frequency of licking the second kid and the frequency of suckling (0.6) the second kid (r=0.57, 0.6, P=0.1 respectively). We concluded that placentophagia could be a highly motivated behavior in post-parturient goats. Grant UNAM-DGAPA-PAPIIT IN217012.

In Romanov ewes the maternal experience affects the establishment of recognition of the lamb

García Y. González, Ethel, Cuellar, Alfredo, Nandayapa, Edith, De La Cruz, Héctor, Ayala, Karen and Terrazas, Angélica, Facultad de Estudios Superiores Cuautitlán, U.N.A.M., Departamento de Ciencias Pecuarias y de Ciencias Biológicas, Km 2.5 Cuautitlán-Teoloyucan, San Sebastian Xhala, Cuautitlán Izcalli, Edo. Mex., 54714, Mexico; eth_cat@hotmail.com

We assess if maternal experience impairs the ability of the mother to recognize its lamb in Romanov ewes. Eight primiparous and 7 multiparous ewes were used. Maternal olfactory recognition of the lamb was assessed at 4 h p.p. by testing selective nursing behavior. Non-olfactory recognition was assessed at 8 h in a two-choice test excluding olfactory cues. A Mann-Whitney test was to compare between groups and Wilcoxon test for comparing within groups. At 4 h multiparous tend to accept more at the udder the own lamb than primiparous (3.4±0.71 vs. 1.5±0.88, P=0.07). Primiparous tend to emit less maternal bleats with own lamb than multiparous (12.5±4.2 vs., 20.8±3.7, P=0.08), primiparous emitted more high pitch bleats with own lamb than multiparous (13.5±4.65 vs. 4±1.29, P=0.04). Within groups primiparous to permit to alien lamb stay longer time near the udder (P=0.06). Non-significant difference were found in the frequency of rejection and the acceptance to the udder between the own and alien lamb (P>0.1). While in multiparous either the time or the frequency to accept to the udder was higher for own lamb than alien (P<0.05). Multiparous emit more low pitch bleats with own lamb than alien (P=0.02). The frequency of aggressive behavior was higher toward alien than own lamb (<0.05) in both groups. At 8 h only multiparous ewes show clear preference for their own lamb, in order that they spend longer time near to and looking toward the own lamb than to the alien. Also they visit them more frequently (P<0.05). While primiparous did not show the same ability (P>0.05). In conclusion in Romanov ewes the lack of maternal experience impairs either olfactory or non-olfactory recognition of their lambs. Grant UNAM-DGAPA-PAPIIT IN217012.

Behavioral changes in grazing lambs provoked by their separation from dam at weaning

Damián, J.P.[1], Ramírez, M.J.[2], Hötzel, M.J.[3], Banchero, G.[4] and Ungerfeld, R.[2], [1]Facultad de Veterinaria, Universidad de la República, Uruguay, Departamento de Biología Molecular y Celular, Lasplaces 1550, 11600, Montevideo, Uruguay, [2]Facultad de Veterinaria, Universidad de la República, Uruguay, Departamento de Fisiología, Lasplaces 1550, 11600, Montevideo, Uruguay, [3]Universidade Federal de Santa Catarina, Florianópolis, SC, Brazil, Laboratório de Etologia Aplicada e Bem-Estar Animal, Departamento de Zootecnia e Desenvolvimento Rur, Rod Admar Gonzaga 1346, 88034001, Florianopolis, Brazil, [4]INIA 'La Estanzuela', Colonia, Uruguay, Unidad de Ovinos, Ruta 50, Km 11, Colonia, 70000 Colonia, Uruguay; jpablodamian@gmail.com

The study aimed to determine the influence of the separation from their dams on the behavioural response at weaning in grazing lambs. Fourteen lambs were separated from their dams at 24 h of age and artificially reared (AR); they received sheep's milk from teats in similar amount and frequencies as suckling lambs, which were kept with adult ewes, and had minimal contact with humans. Another 13 lambs remained with their dams since birth (DR). All lambs received solid supplement from 20 d of age, grazed improved pastures, and had free access to water. At 75 d of age (d0), DR lambs were separated from their dams and AR lambs stopped receiving milk. Behaviours were recorded every 10 min for 6 h per day, from d-3 to d4. Vocalizations were recorded during 30 s every 10 min using 0/1 sampling. Percentage of observations of each behaviour was analysed with a mixed model. Frequency of observations in which lambs were walking, vocalizing and pacing increased from d0 to d2 ($P<0.0001$). The frequency of grazing decreased on d0 ($P<0.0001$), then increased to the maximum frequency on d2 ($P<0.0001$), and returned to initial values on the d3. Frequency of playing decreased to a minimum on d2 ($P<0.0001$), remaining low until d4. Walking, vocalizing and pacing frequencies on d0 were greater in DR than AR [walking: 18.2±2.7 vs. 7.9±1.1 ($P<0.0001$); vocalizing: 39.7±3.0 vs. 5.5±1.5 ($P<0.0001$); pacing: 9.3±2.6 vs. 0.0±0.0 ($P<0.0001$); for DR and AR respectively]. Differences were still present on d1 ($P<0.05$). The decrease of frequency of grazing on d0 was greater in DR (45.1±3.0) than AR (63.5±1.9) ($P<0.0001$) lambs, and remained lower on d2 and d4 ($P<0.05$). Drinking water and playing were not influenced by treatment. Grazing, walking, vocalizing and pacing may be behaviours mainly induced by the separation from their dam at weaning.

Familial separation and transport stress in a family of captive snow leopards (*Panthera uncia*)

Owen, Yvonne[1], Amory, Jonathan[1], De Luna, Carlos J.[1] and Melfi, Vicky[2], [1]Writtle College, School of Sport, Equine and Animal Science, Chelmsford, Essex, CM1 3RR, United Kingdom, [2]Taronga Conservation Society, Taronga Zoo Park, Bradley's Head Road, Mosman, NSW, 13 1314, Australia; yvonne.owen@writtle.ac.uk

The transportation of animals, although a potential stressor, is an important aspect of animal management in terms of breeding programmes and conservation. However, stressors are known to challenge homeostasis and in extreme cases can lead to behavioural, reproductive and/or pathological problems. This study examined the effect of separating a juvenile snow leopard (Panthera uncia) from her sire and dam at Paradise Wildlife Park, Broxbourne, UK and transporting her by road over 1,400 km to Kittenberger Zoo, Hungary. The study also examined the effect of familial separation on the family as a whole. Baseline measures of animal behaviour plus corticosterone concentrations were taken four months prior to transportation and based on keeper trait ratings and eighteen days of faecal corticosterone concentrations (FCCs) respectively. Following transportation, keepers were interviewed regarding any behavioural changes and a further 18 days' worth of FCCs collections was made. Post transportation behaviour of the sire appeared unchanged; however the dam immediately exhibited reduced activity and appetite suppression. Behavioural changes were most apparent in the transported juvenile with a substantial decrease in activity and appetite suppression for a period of three months. Nine days post-transportation the juvenile exhibited a maximal 256% increase from baseline FCCs (80.8 ng/g + 41.51 ng/g), with an average increase of 122% over the 18 day sample period. Results for the sire indicated an 8% decrease from baseline FCCs (51.4 ng/g + 30.63 ng/g), and for the dam a 4% decrease from baseline FCCs (106.7 ng/g + 75.79 ng/g). It is unclear what effect, if any, familial separation had on the sire and dam. However, separation and transportation are both likely causes of stress, manifested as extended behavioural changes and FCC changes for the juvenile snow leopard. These results should be considered when visiting best practice guidelines for the relocation of animals.

An analysis of subgroup formation and behavioural synchronization in a herd of Shetland pony mares

Hauschildt, Verena and Gerken, Martina, University of Göttingen, Department of Animal Sciences, Albrecht-Thaer-Weg 3, 37075 Göttingen, Germany; vhausch@gwdg.de

Social structure and behavioural synchronization were examined in a group of 10 Shetland pony mares kept on pasture. Direct field observations covered 1 month, lasting 2 h each between 08:00 and 20:00 on 35 occasions. Nearest neighbour identity and distance were recorded at 15 min intervals by point sampling, while social interactions were recorded continuously. Individual rank was derived from outcomes of agonistic interactions. Age, coat colour, body weight, wither height and body condition scores (BCS) were assessed as possible determinants of social structure. Relationships between traits were analysed using Kendall's τ correlations. All mares preferred certain individuals as neighbours, and identities of frequent neighbours correlated with frequent grooming partners ($P<0.05$). Rank, age, height, and weight neither correlated between most frequent neighbours nor grooming partners ($P>0.05$). However, the correlation between rank position and frequency of being closest (<1 m) to the nearest neighbour ($\tau_b=0.56$; $P=0.025$) indicated that high-ranking mares associated more strongly with other mares. Frequency of mutual grooming was unaffected by rank ($\tau_b=0.38$; $P=0.128$), but tended to be directed towards lower-ranking individuals. Rank was not correlated with age, height, weight, or BCS ($P>0.05$). Multivariate analyses of both frequency of neighbourship (non-metric multidimensional scaling) and mean distances between nearest neighbours (hierarchical cluster analysis) revealed 2 subgroups consisting of 3 or 4 mares, respectively. Unexpectedly, mares within one subgroup were dissimilar in external appearance and rank. The mares synchronized their behaviour about 24% of the observation time, mostly related to feeding behaviour. Higher behavioural synchronization was observed between neighbours (mean ± SD, 81.00±2.96%) and within both subgroups (50.64% and 51.61%, respectively). The results of this study suggest that behavioural synchronization is a suitable indicator of social cohesion in grazing horses. In group housing of horses, subgroup formation may occur and should be considered in design of housing facilities.

Impact of unfamiliar adults on the behavior of weanlings

Henry, Séverine[1], Zanella, Adroaldo[2] and Hausberger, Martine[1,3], [1]University Rennes 1 – UMR CNRS 6552 EthoS, Station Biologique de Paimpont, 35380 PAIMPONT, France, [2]Scottish Agricultural College, Roslin Institute Buildinbg, Easter Bush, EH25 9RG Midlothian, United Kingdom, [3]University Rennes 1 – UMR CNRS 6552 EthoS, 263 av. Génréral Leclerc, Campus de Beaulieu, 35042 Rennes, France; severine.henry@univ-rennes1.fr

Different approaches have been investigated to reduce weaning stress in foals. The present study examines whether the presence of unrelated adult horses at weaning would reduce the social stress of weaning and the emergence of undesirable behaviours. This study was conducted in two sites. Subjects were 12 colts and 20 fillies from Arab (site 1), Anglo-Arab and French-Saddlebred (site 2) breeds. Weaning occurred outdoors at 4.5 (site 1) or 7 (site 2) months of age. At both sites, foals were divided into Peer-Weaned (PW) and Adult-Weaned (AW) groups: in the formers, foals were maintained after weaning in same-age groups; in the latters, two unfamiliar adult horses were introduced following the removal of all mares. At farm 1, one PW group (n_{PW1-1}=5) and one AW group (n_{AW1-1}=4) were constituted. At farm 2, four balanced groups, 2 PW groups (n_{PW2-1}=6, n_{PW2-2}=6) and 2 AW groups (n_{AW2-1}=6, n_{AW2-2}=5), were formed. Thus, three separate trials were conducted, with each trial consisting of one AW group and one PW group. PW and AW groups were kept in separate pastures throughout the study period. Levels of salivary cortisol were measured one day before to four days after the weaning day. Foals behaviours were recorded prior to weaning and through one month post-weaning. Statistical significance was tested using non-parametric tests. In all groups, weaning was followed by increased vocalization, increased locomotion and increased salivary cortisol concentration. However, signs of stress were less pronounced and shorter in duration in weanlings with adults (e.g. whinnies: $P<0.05$; salivary cortisol: $P<0.05$). Only foals without adults exhibited for several weeks post-weaning aggressiveness towards peers ($P<0.05$) and abnormal behaviours ($P<0.05$) including excessive wood-chewing and redirected sucking towards peers. In conclusion, introducing adults to minimize weaning stress in foals and later on aggressiveness and abnormal behaviours appears as the most promising approach to date.

Sow nursing synchronization in modified and unmodified visual and acoustic contact

Šilerová, Jitka, Špinka, Marek and Neuhauserová, Kristýna, Institute of Animal Science, Department of Ethology, Přátelství 815, 104 00 Prague, Czech Republic; jsilerova@atlas.cz

The nursings of sows housed in one room tend to be synchronized. Previously we documented that the nursing synchronization was rather high in sows housed within one room. The vocal and visual stimuli impact on nursing synchronization was investigated through manipulation of the stimuli input; the nursing synchronization among sows with unmodified contact (NFS) and the focal sows was examined. We assumed: H1. Auditory contact is adequate for nursing synchronisation to develop. H2. Nursing frequency is increased by synchronization. Ten triplets of focal sows were housed in the 3 treatments at the same time within the room of 14 pens (C and S): Isolation (I) – visually, olfactory and auditory isolated room; Sound (S) – visually isolated pen; Control (C) – pen with the visual, olfactory and acoustic contact. The NFS were housed in the remaining farrowing pens within the room of 14 pens. The nursing behaviour was observed for 6 hours at 7 and 21 days post-partum: (1) through direct interval sampling of all lactating sows; and (2) via three focal sows video recording. Nursings of at least two of the sows that began within 3 min of each other were considered to be synchronized. The NFS were more synchronized with C than with S sows ($t_{(1,40)}$=2.26, LSmeans=0.2644, $P<0.05$). Both the C and S sows were much more synchronized with the NFS than the I sows ($t_{(1,40)}$=5.48, LSmeans(C)=0.6408, $P<0.0001$) ($t_{(1,40)}$=3.22, LSmeans(S)=0.3765, $P<0.01$). The interval before synchronized nursing was not shorter than such before non-synchronized nursing for the C and S sows ($F_{(1, 247)}$=14.37, $P<0.001$, LSmeans: synchronized=52.41 min, non-synchronized=44.12 min). In conclusion the auditory contact was sufficient for nursing synchronisation to develop (although olfactory cues could contribute). Nursing frequency was not increased by nursing synchronization.

Social recognition in weaner pigs after two weeks of separation

Böttcher, Uta M. and Hoy, Steffen, Justus Liebig University, Department of Animal Breeding and Genetics, Bismarckstrasse 14, 35390 Giessen, Germany;
Uta.M.Boettcher@vetmed.uni-giessen.de

First encounters of unacquainted pigs result in vigorous fighting. In stable groups only few fights occur. Our objective was to study the extent of social recognition in littermates by analysing the agonistic interactions and body lesions during social encounters after separation. 228 weaner pigs out of 26 litters were included in the study. After a suckling period of four weeks the piglets were weaned and placed in a raising unit. Two same-aged litters were each mixed either at weaning or after an additional period of three weeks staying as complete litter and placed into two different pens (A and B). After another 14 days (at an age of 6 or 9 weeks respectively) one pig of pen A was paired with one pig of pen B and they were confronted in a test arena for 45 minutes. Each pair consisted of either former littermates or unacquainted pigs. The encounters were videotaped and analysed focusing on agonistic behaviour. After the encounter each pig was examined for skin injuries. Due to the distribution of the data non-parametric methods were used. 45.6% of all pairs fought (total duration in minutes: 3:31±6:36; median, min-max: 0:00, 0:00-33:52; 75th percentile: 0:32). In case of 6 week old pigs no influence of the level of familiarity on any measured parameter was found. In case of 9 week old pigs 13.3% of the former littermates and 42.3% of the unacquainted pairs fought ($P<0.05$). The fights occurred significantly later in littermates (997 s vs. 567 s; $P<0.05$). Only 3.3% of the littermates showed relevant skin lesions whereas in unacquainted pairs this frequency was significantly higher (38.5%; $P<0.001$). Social recognition after a separation of two weeks seems to exist for the 9 week old pigs, but not for the 6 week old pigs. Age or further stabilization of social bonds after weaning might explain this.

Behavioural response in dairy cows with and without calf-contact to calf-odour in the parlour

Zipp, Katharina[1], Barth, Kerstin[2] and Knierim, Ute[1], [1]University of Kassel, Farm Animal Behaviour and Husbandry Section, Nordbahnhofstr. 1a, 37213 Witzenhausen, Germany, [2]Johann Heinrich von Thünen-Institut, Federal Research Institute for Rural Areas, Forestry and Fisheries, Trenthorst 32, 23847 Westerau, Germany; zipp@uni-kassel.de

Combined dairy suckling and milking systems provide some advantages, especially for the mother-reared calves, but pose milk let-down problems during machine-milking. One potential, yet up to now unsuccessful approach to improve milkability is to stimulate the cows by calf-odour. This study aimed to ascertain whether such odour stimuli are perceived and produce a differentiated response depending on cow-calf contact and odour identity. 18 cows, 5 heifers, separated from their calves after birth ('control'), and 17 cows, 6 heifers with permanent calf contact ('contact') were tested in their third lactation week. They were of the breeds Holstein-Friesian and German Red Pied. During six of twice daily milkings, in the milking parlour they were presented with two of altogether three stimuli (bag with hair of the own calf=Co, with hair of an alien calf=Ca, without hair=0) in two punnets on top of the other in randomised order, resulting in all possible combinations and positions of each stimulus. Duration of sniffing and licking at the punnets/duration of stimulus presentation (=response) was recorded for 3 minutes after stimulus presentation by continuous focal animal sampling from videotapes. More heifers than multiparous cows (91% vs. 51%) reacted at least once in all trials to any sample ($\chi^2(1)=5.48$, P=0.019). Additionally, in tendency more 'contact'-cows responded (78% vs. 44%; $\chi^2(1)=3.29$, P=0.070). Within these responsive cows, no difference between 'contact'- and 'control'-cows in their response to Co+Ca vs. 0 (difference (Co+Ca)-0; P=0.282) or Co vs. Ca (P=0.128, Mann-Whitney-test) was found. However, they in general responded longer to Co (median: 3.98%) than to Ca (2.06%) and to 0 (0.00%; P=0.004, Friedman's ANOVA). The relatively low response rates hint at a certain inappropriateness of stimulus presentation. However, the responsive cows indicated that they were able to perceive different odours and were most interested in odour from their own calf.

Goats distinguish between group members and non-members even when head cues are missing

Keil, Nina M.[1], Imfeld-Mueller, Sabrina[2], Aschwanden, Janine[1] and Wechsler, Beat[1], [1]Swiss Federal Veterinary Office, Centre for Proper Housing of Ruminants and Pigs, Agroscope Reckenholz-Taenikon ART Research Station, 8356 Ettenhausen, Switzerland, [2]ETH Zurich, Institute of Agricultural Sciences, Animal Behaviour, Health and Welfare Unit, Universitaetstrasse 2, 8092 Zurich, Switzerland; nina.keil@art.admin.ch

In this study, we investigated whether goats can distinguish a group member from a non-member, even when the head of the goat in question cannot be seen. A total of 45 adult female goats ('walkers') were trained to walk along a passageway at the end of which they expected food ('trial run'). While walking, they passed by another female goat ('test goat') whose trunk and hind legs alone were visible. Using 19 individuals, ten pairs of test goats consisting of one goat from the walker's and one from a different group were matched in terms of body size, constitution, colour and coat length. The walkers completed two, four or six trial runs depending on the number of pairs suitable for a given walker. Data of 109 trial runs were analysed by using generalised linear mixed-effects models with crossed random effects. Explanatory variables were whether the test goat was from the walker's group (yes/no) and whether the test goat bleated during the trial run (yes/no). On average (±SD), the walker spent a total of 8.7±6.4 s exploring the test goat visually and olfactorily if the latter was from a different group, and only about half as long (4.2±3.7 s) if it was from her own ($P<0.001$). In particular, the length of time a walker spent observing a test goat whilst approaching the latter was significantly longer if the test goat belonged to a different group (2.5±1.6 vs. 1.4±1.5 s; $P<0.001$). A test goat from a different group was also sniffed at markedly longer than one from the same group (3.4±4.6 vs. 0.9±1.9 s; $P=0.001$). In conclusion, these results suggest that goats easily discriminate between group members and non-members, even when head cues are missing. Olfactory as well as visual cues are probably important for identifying group members.

Identification of morphological parameters associated with affiliative patterns to enhance sociopositive herd structures: potential relevance of beards in dairy goats

Schaefer, Sybille, Schuster, K. and Wasilewski, Anja, Philipps-Universität Marburg, Biologie – Spezielle Zoologie, Karl-von-Frisch-Str. 8, 35032 Marburg, Germany; wasilews@staff.uni-marburg.de

Identifying morphological parameters reliably recognised by humans and associated with affiliative herd patterns may help enhance sociopositive structures under commercial husbandry conditions. Spontaneous behaviour of 16 dairy-goats (Thüringerwaldziege), varying in age and appearance (coat-colour, horn-width, possession/lack of wattles and/or beard), was recorded under loose-housing conditions. For two 5-week observation periods (31 d apart), data were collected a minimum of 3 h/d on 5 d/week. Nearest-neighbour frequencies (instantaneous-scan-sampling: 2-min-intervals between focal animals, 60-min intervals between consecutive scans, min. 2 recordings/animal×day; total = 53 h 20 min) and agonistic interactions (behaviour-sampling: additionally ≥2 h/day; total = 137 h 45 min, 4,709 interactions) were recorded. A synthesis of multidimensional scaling (MDS) and cluster analysis served to visualise association structures generated from neighbour frequencies. Dominance hierarchy was derived by calculating dominance indices (DI). Linearity was quantified by Landau's h (0.77). DI ranged from 5% to 86%. Synthesis diagrams grouped 10 goats into closer, invariably pair-wise, associations. Association structures did not reflect dominance relationships: Pairs included animals similar in dominance index as well as those differing widely (DI-difference: min.=4%, max.=44%). Comparing association structures with the 5 independent individual characteristics (age, colour, horns, wattles, beard) did not yield consistent results for the first four. Concerning beards, synthesis diagrams revealed non-bearded goats (5 animals) not to be randomly distributed among bearded goats. 'Pairings' occurred either between bearded (3 pairs) or non-bearded individuals (2 pairs). No dyadic patterns were detected for 1 non-bearded and 5 bearded goats. Given the group size and preordained distribution of characteristics among the herd members examined, further research conducted within a balanced study design appears worthwhile, as the influence of physiognomic characteristics on association patterns has previously been demonstrated for horned vs. hornless rams. Behaviour patterns could be linked to neurological correlates. Meanwhile, the tentative recommendation for husbandry practice is to stock even numbers of beardless goats whenever possible.

Living in high animal density group: more frequent agonistic interactions but not positive social behaviour in goats

Vas, Judit, Andersen, Inger L. and Chojnacki, Rachel, Norwegian University of Life Sciences, Department of Animal and Aquacultural Sciences, P.O. Box 5003, 1432 Aas, Norway; judit_banfine.vas@umb.no

Goats are kept at highly different densities across the world, but little is known about the effect of the variability in space available to each individual. Therefore we studied the impact of three different animal densities on social interactions and stress responses of goats during pregnancy. Fifty-four Norwegian dairy goats were divided into pens with low, medium or high density (3, 2 and 1 m^2 per goat, six animals in each pen, three groups in each treatment) from early stage of pregnancy until parturition. Direct, continuous observations of social interactions were made (3 hours per observation) and the goats' weight, body condition and serum cortisol (on two consecutive days) were measured on three occasions (one week after the beginning of the treatment, and in the second and last third of pregnancy). The animals' rank was calculated based on displacement events (observed in dyads, using Noldus MatMan). Data were analyzed using generalized linear mixed models and Spearman rank correlations in SPSS. According to our results offensive behavior was shown more frequently in goats kept at high density than in goats kept at medium or low density (P=0.011). There was no difference between treatment groups in the number of positive interactions. The observed social hierarchy was non-linear and individual rank order tended to be stable throughout the treatment period. Individual rank order correlated weakly with weight (rho=0.41) and stronger with offensive behavior initiated by the animal (rho=0.67). The results of the cortisol analysis will be presented at the meeting. In conclusion, the more frequent agonistic interactions in goats kept at 1 m^2 per animal shows that in goats negative, but not positive social interactions are density dependent.

The effect of order of ewes entry into parlour on their milkability and milk composition

Antonic, Jan[1], Macuhova, Lucia[2], Uhrincat, Michal[2], Jackuliakova, Lucia[1] and Tancin, Vladimir[1,2], [1]Slovak University of Agriculture, Tr.A. Hlinku 2, 94901 Nitra, Slovakia, [2]Animal Production Research Centre Nitra, Hlohovecka 2, 95141 Luzianky, Slovakia; tancin@cvzv.sk

The aim of our investigation was to evaluate the effect of order of entry into milking parlour on the milkability parameters and milk composition. We used 353 lactating dairy ewes of three breeds and their crossbreeds Tsigaj, Improved Valachian and Lacaune. They were milked in 1×24 stalls milking parlour and their order was recorded during four continuous evening milkings two months after lamb weaning. On the fifth day the experimental flock of ewes were created before evening milking. Within the flock 24 ewes (First group – 1st always milked within first two turns) and 24 ones (Second group – 2nd always milked within last three turns) were measured. Milk flow kinetics was recorded and milk samples were collected during evening milking. During pre-exp. and during exp. measurements the ewes had a free choice to enter the parlour. Statistical evaluation was done by T-test of independent samples of Statistica program (version 8.0, StatSoft. Inc.) Total milk yield (MY), machine MY, and maximal milk flow rate were significantly higher in 1st group ($P<0.05$). Also, milk flow latency and time of second emission of milk flow were in 1st group shorter than in 2nd group. In the 1st group were recorded 50.9% of two emissions of milk flow as compared with 27.6% in 2nd group resp.. Ewes in 1st group had a tendency to be older than ewes in 2nd group. In milk composition difference between groups was found out for the content of fat-free dry matter ($P<0.05$) only. In conclusion we could demonstrate the effect of order of ewe entry into parlour on milk yield and on milk flow kinetic. Changes in milk flow kinetic between groups could indicate different adaptability of ewes to milking and therefore probably their willing to be milked out.

Social support alters behavioural and neuroendocrine responses to a single social isolation in juvenile pigs

Hameister, Theresa[1], Kanitz, Ellen[1], Tuchscherer, Armin[2], Tuchscherer, Margret[1] and Puppe, Birger[1], [1]Leibniz Institute for Farm Animal Biology, Behavioural Physiology, Wilhelm-Stahl-Allee 2, 18196 Dummerstorf, Germany, [2]Leibniz Institute for Farm Animal Biology, Genetics and Biometry, Wilhelm-Stahl-Allee 2, 18196 Dummerstorf, Germany; hameister@fbn-dummerstorf.de

Social isolation is a commonly used model of psychosocial stress inducing a wide range of adaptation mechanisms. We investigate if and how social support in neonatal pigs can buffer the ethological and physiological consequences of a 4 h social isolation in an unfamiliar environment (isolation box). Piglets were classified into four treatment groups (control without isolation, isolation alone, isolation with familiar or unfamiliar piglets of the same age) and examined at 7, 21 or 35 days of age. The behavioural responses were analysed during the isolation procedure and in repeated open-field/novel-object (OF/NO) tests. Blood samples were taken before and after the treatment for hormone analyses. Thereafter, some piglets were euthanized and their brains were prepared for the analyses of stress-related gene expression. Data were evaluated by ANOVA and pairwise comparisons using SAS (procedure MIXED). Piglets isolated alone were more excited in the OF/NO test (locomotion: $P<0.05$; escape attempts: $P<0.01$) and showed higher cortisol concentrations ($P<0.001$) compared to controls. These piglets displayed a lower ratio of mineralocorticoid (MR) and glucocorticoid receptor (GR) in different brain areas ($P<0.05$). Social support had an overall calming effect on behavioural responses of piglets both during isolation and OF/NO tests compared to piglets isolated alone. The cortisol concentration was reduced ($P<0.1$) and the imbalance of MR/GR in the brain was reversed ($P<0.05$). Moreover, piglets isolated with a familiar conspecific were more passive in their behavioural test responses than piglets with an unfamiliar one. The 7-day-old piglets showed higher behavioural activity and arousal during the test situations compared to the older piglets ($P<0.05$). The results suggest that negative consequences of psychosocial stress could be buffered by social support probably leading to improved welfare and an enhanced ability to cope with challenging situations. In this process, the familiarity on individuals plays a modifying role and should be considered.

Relationships of social behaviors with neonatal plasma oxytocin in heifer calves

Yayou, Ken-ichi[1], Ito, Shuichi[2] and Yamamoto, Naoyuki[3], [1]National Institute of Agrobiological Sciences, Division of Animal Sciences, 2, Ikenodai, Tsukuba, 305-8602, Japan, [2]Tokai University, Department of Agriculture, Kawayo, Minami Aso-mura, Aso-gun, 869-1404, Japan, [3]Western Region Agricultural Research Center, Livestock Production and Wildlife Management Research Division, Ooda, 694-0013, Japan; ken318@affrc.go.jp

During the neonatal period, oxytocin is reported to have organizational effects on central nervous system, which have long-lasting effects on social behaviour. Thus, we examined the relationships between neonatal oxytocin level and social behavioural traits in cattle. Jugular blood oxytocin concentrations at resting conditions were measured in 20 Holstein heifer calves at 1 (OXT1), 2 (OXT2), and 6 (OXT6) weeks of age. Social behaviours were observed at about 6 weeks of age for two hours each on the first and the second day after the first introduction into a group of 4 peers. During 10 to 14 months of age, social behaviours were observed for 6 hours each, immediately after and 1, 4, 20 weeks after the introduction into a group of 20 heifers. Relationships of oxytocin concentrations with social behavioural traits in each observation were analyzed using principal components analysis and Spearman rank-correlation coefficient. At about 6 weeks of age, OXT6 tended to correlate positively with 'negative coping trend (e.g. less social exploration)' on the first day (r_S=0.386, $P<0.1$) and 'affiliative trend (e.g. more affiliative behaviour)' on the second day (r_S=0.400, $P<0.1$). During 10 to 14 months of age, OXT1 and OXT2 significantly and positively correlated with 'alienated from peers (e.g. longer individual distance)' 1week after (r_S=0.457 and 0.450, $P<0.05$), and with 'unilateral affiliative trend (e.g. more affiliative behaviour to peers but less from peers)' 4 weeks after (r_S=0.620 and 0.634, $P<0.005$), and negatively correlated with 'less social interaction' 20 weeks after the introduction into the group (r_S=-0.412, $P<0.1$ and -0.549, $P<0.05$). These results likely reflect some influence of neonatal oxytocin level on the individual differences in social behavioural traits in cattle.

Body language of dogs responding to different types of stimuli

Norling, Yezica[1], Wiss, Viktoria[1], Gorjanc, Gregor[2] and Keeling, Linda J.[1], [1]Swedish University of Agricultural Sciences, Department of Animal Environment and Health, PO 7068, 751 41, Uppsala, Sweden, [2]University of Ljubljana, Animal Science Department, Groblje 3, 1230 Domzale, Slovenia; Yezica.Norling@slu.se

The aim of this study was to identify behavioural indicators of positive emotional states in 9 female beagles by studying a variety of subtle movements. Dogs were trained to remain stationary in a test cubicle with transparent sides while filmed from the front, side and above. Dogs faced a stimuli presentation device with a shutter that revealed one of three stimuli; a familiar person smiling and talking in a praising voice (P1), a meatball (P2), both presumed positive, and a wooden block (N) presumed neutral. To each dog, each stimulus was presented once in each of 6 sequences. Behaviour was recorded for 5 seconds before and after each stimulus presentation. Data were analysed with generalized linear mixed model. Only significant differences with $P<0.05$ from Tukey post-hoc test are reported. As expected, there were no behavioural differences before the stimuli appeared. The initial reaction (1-2 sec) to all stimuli was to raise the head and turn it briefly to the right. Subsequently (sec 3-5), the most substantial changes were elicited by the familiar person with greater amplitude of tail wags (smallest difference P1: 4.07±0.95 vs. P2: 1.56±0.95 (mean no of zones±SE) in sec 4; 1 zone=13° deviation from centered position) and elevation of the tail (smallest difference P1: 2.15±0.30 vs. N: 1.16±0.20 in sec 5; one zone= 20°) compared to all other stimuli. Also paw lifting (P1: 0.19±0.12 vs. P2: 0.08±0.05 (mean lifts/s±SE) in sec 3) and lip licking were performed more often (least difference P1: 0.24±0.08 vs. N: 0.05±0.04 (licks/s±SE) in sec 4) to the familiar person. Although we cannot confirm the positive valence of the emotional states in this first study, these results lay the foundation for further controlled studies of behaviour or combinations of behaviour in order to map the body language of emotional states.

Owner-perceived behavioural indicators of positive affective states in companion dogs

Buckland, Emma, Volk, Holger, Burn, Charlotte and Abeyesinghe, Siobhan, The Royal Veterinary College, Hawkshead Lane, North Mymms, Hatfield, Hertfordshire, AL9 7TA, United Kingdom; elbuckland@rvc.ac.uk

Owners are important proxies for veterinary, welfare and quality of life assessments; accurate interpretation of pet behaviour is therefore necessary. An online survey of UK and Ireland dog owners (n=447) examined descriptive behavioural signs that owners used to identify canine positive affect (CPA), for both high and low arousal states. Effects of owner and dog variables on the citation of specific behaviours were examined using multivariate binary logistic regression; all significant influences (P<0.05) on preceding behaviour are indicated in parentheses. Owners described a wide range of CPA indicators. The most frequently (>20% respondents) reported indicators of high arousal CPA were: barking; active; playful; 'wriggly' and head held high (both decreasing with dog age); alert; tail wagging fast (great in Hound & Toy Kennel Club registered breeds than crossbreeds) or held high; mouth open; 'smiling' or panting; eyes wide open or bright; and ears pricked up (greater in Gundog, Utility & Toy breeds than crossbreeds). Low arousal CPA was most frequently (>20% respondents) described by: sighing/huffing (increasing with more dogs owned); no vocalisations, head resting and ears relaxed (for all three, greater when respondents were owners only, i.e. did not also work with or breed dogs); calm; affectionate; sleepy; lying; mouth closed/loose; eyes slightly closed; tail low/ relaxed. Overall, owner-reported signals of CPA varied greatly, particularly across levels of owner experience. Low arousal states were less well described, suggesting they may be less well understood and/or difficult to interpret. CPA expression may differ with dog-related factors (e.g. individual, breed, age), and/or anthropomorphic interpretation of such characteristics may constrain accurate human identification. Thus further study will examine (1) variation in CPA expression; and (2) validate owners' abilities to accurately identify context-specific CPA against objective behavioural and physiological indicators.

Does repeated exposure to positive events alleviate pessimistic-like judgment in chronic-stressed sheep?

Destrez, Alexandra[1], Deiss, Véronique[1], Leterrier, Christine[2], Calandreau, Ludovic[2], Lévy, Frédéric[2] and Boissy, Alain[1], [1]INRA, UMR 1213 Herbivores, ACS, 63122 Saint-Genès Champanelle, France, [2]INRA, UMR 85 PRC, 37380 Nouzilly, France; alexandra.destrez@clermont.inra.fr

Chronic stress induces detrimental behavioural effects in animals, e.g. by inducing a pessimistic-like environment perception. On the contrary, enriched environments can elicit optimistic-like perception. We investigated whether repeated exposure to various positive events could alleviate pessimistic-like judgment in chronically stressed sheep. Fifteen lambs (chronic-stressed group) were submitted, over 6 weeks, to a treatment known to induce chronic stress (i.e. exposure to various unpredictable, uncontrollable and aversive events every day and/or night). Fifteen other lambs (REPE group) were submitted to the same treatment designed to induce chronic stress plus they were exposed to various positive events, i.e. wool brush, positive tactile contacts with humans and food anticipation following a light announcing daily food distribution, throughout the 3 last weeks of treatment. Finally, fifteen other lambs were housed in the same standard breeding conditions as the chronic-stressed group and exposed to positive tactile contacts with humans over 6 weeks (controls). Before treatment, all lambs were trained to approach/avoid a food bucket depending on its location (go/no-go procedure). After treatment, lambs were individually exposed to a judgment test consisting of presenting the bucket in intermediate locations (i.e. ambiguous situations) and the latency to approach the bucket was recorded. During the judgment test, sheep of the REPE group approached the ambiguous buckets faster than chronic-stressed sheep (3.0±0.3 s vs. 6.0±1.6 s, Kruskal-Wallis test K=8.7, with Bonferroni corrections P=0.01) and controls approached the ambiguous buckets faster than chronic-stressed sheep (3.2±0.2 s vs. 6.0±1.6 s, Kruskal-Wallis test K=8.7, with Bonferroni corrections P=0.05). Therefore chronic-stressed sheep experiencing repeated exposures to positive events, as well as controls, demonstrated a more optimistic-like judgment than chronic-stressed sheep. Further studies are necessary to assess and/or improve such a positive treatment as a method to alleviate negative effects of chronic stress.

Play behaviour of dairy calves after weaning according to grain intake

Miguel-Pacheco, Giuliana G.[1,2], Vaughan, Alison[2], Rushen, Jeffrey[2] and De Passillé, Anne M.[2], [1]University of Nottingham, School of Veterinary Science and Medicine, Sutton Bonington, Loughborough, LE12 5RD, United Kingdom, [2]Pacific Agri-Food Research Centre, Agriculture and Agri-Food Canada, 6947 Highway 7, P.O. Box 1000, Agassiz, BC, V0M 1A0, Canada; svxgm@nottingham.ac.uk

We examined effects of age and weaning according to grain intake on play running in 55 female Holstein calves (6±1 days of age) housed in groups of 4 or 8 in 7.08×4.74 m pens, with expanded metal floor and sawdust bedded area. Calves had free access to milk, grain, and hay from automated feeders. Calves were observed for 15 hours/day (between 08:00 and 23:00) at three ages relative to weaning (with the age of weaning depending on each individual calf's grain intake): Before Weaning: 2 of the 5 days prior to the start of weaning; Post-weaning period 1: 2 of the 3 days immediately after weaning completed; Post-weaning 2: 2 days between the 7th and 10th day after weaning completed. Running was scored from videos and defined as rapid-forward movement that may include jumping, bucking and/or kicking (one or two legs), lasting 3 s minimum. Before weaning there was no significant correlation between digestible energy (DE) intake (calculated from milk, grain and hay intakes; Mean: 7.96 MJ/day, SD 2.56) and the duration of running: r_s=0.11, P=0.44; or running frequency: r_s=0.11, P=0.479. Running duration was lower in Post-weaning 1 compared to before weaning (4.9±0.7 s/d versus 61.6±8 s/d; P<0.001). In Post-weaning 1, there was a tendency for a low positive correlation between DE intake (Mean: 11.23 MJ/day SD 3.11) and running duration (r_s=0.24, P=0.072) and a low significant positive correlation with running frequency (r_s=0.29, P=0.030). Before weaning, older calves ran less than younger ones (RD: r_s=-0.39, P<0.05 and RF: r_s=-0.41, P<0.05). Dairy calves reduced their locomotor play after weaning and as they aged. After weaning, play was reduced according to the degree of reduction in energy intake. Play behaviour may be a useful tool to measure the impact of management practices in young animals.

Drop in peripheral comb temperature during positive anticipation

Moe, Randi O.[1], Stubsjøen, Solveig M.[1], Bohlin, Jon[2], Flø, Andreas[3] and Bakken, Morten[4], [1]Norwegian School of Veterinary Science, Department of Production Animal Clinical Sciences, P.O. Box 8146 dep, 0033 Oslo, Norway, [2]Norwegian School of Veterinary Science, Department of Food Safety and Infectious Biology, EpiCentre, P.O. Box 8146 dep, 0033 Oslo, Norway, [3]The Norwegian University of Life Sciences, Department of Mathematical Sciences and Technology, P.O. Box 5003, 1432 Aas, Norway, [4]The Norwegian University of Life Sciences, Department of Animal- and Aquacultural Sciences, P.O. Box 5003, 1432 Aas, Norway; randi.moe@nvh.no

A conditioned drop in peripheral skin temperature and concomitant increases in deep body temperature occurs in response to stimuli predictive of and during exposure to unpleasant events in mammals and avians. Thus, a change in peripheral temperature was suggested as a physiological indicator of a negative emotional state. In the present study we investigated peripheral temperature responses during anticipation and consumption of a signaled palatable feed reward, in order to investigate effects of positively valenced emotions on peripheral temperature. Individually housed hens (n=5) previously trained to associate a light signal (CS) with a mealworm reward (US) were during the trial exposed to eight consecutive CS-US pairings repeated every 15 min. The CS-US interval (i.e. the anticipation period) was 25 s. A thermal video camera was used to make continuous recordings throughout the trial; i.e. approximately 135 min. During later video file analysis, comb temperature was recorded every 1 min. A spline regression model (and FFT periodogram) with two interaction terms (including 'hen' and 'trial'; and 'period' (before CS, in the CS-US interval, and idle) and 'temperature', i.e. above or below 30 °C) were used to assess temperature differences for five different hens each participating in the eight CS-US pairings with each trial consisting of 15 time points. On average, comb temperature dropped 1.5 °C (95% CI: ±1.2 °C) immediately after exposure to CS and consumption of reward (P=0.0014) when initial comb temperature was above 30 °C, indicating a peripheral vasoconstriction. This study is the first to show that exposure to a conditioned pleasant event affects thermoregulation in the domestic hen. In conclusion, the use of responses in peripheral skin temperature as an indicator of emotional state and thus, hen welfare, has limitations as it is difficult to interpret emotional valence.

The effect of haloperidol on performance of hens in a Go/NoGo task

Pichová, Katarína[1,2], Horváthová, Mária[2] and Košťál, Ľubor[2], [1]Faculty of Science, Comenius University, Bratislava, Department of Animal Physiology and Ethology, Mlynská dolina, 842 15 Bratislava 4, Slovakia, [2]Institute of Animal Biochemistry and Genetics, Slovak Academy of Sciences, Ivanka pri Dunaji, Moyzesova 61, 900 28 Ivanka pri Dunaji, Slovakia; katarina.pichova@savba.sk

Emotional states occur in response to stimuli that are rewarding or punishing. In humans and animals emotions bias the judgements of future events and ambiguous stimuli. Estimation of cognitive bias in animals is based on operant discrimination task (Go/NoGo) and subsequent testing of affective states effects on prediction of negative or positive events. Dopamine is known to process information about rewards and reward-predicting stimuli. The aim of our experiment was to test the involvement of dopamine in stimulus valuation during Go/NoGo task by treatment with dopamine D2 receptor antagonist haloperidol. Ten adult ISA Brown hens were trained in touch-screen operant chamber the Go/NoGo task, i.e. to peck to a positive stimulus (white circle in 5 hens and 80% grey circle in other 5) associated with a reward (mealworm) and to refrain from pecking at the negative stimulus (80% grey circle in 5 hens and white circle in other 5 hens) to avoid punishment (white noise). After the three Go/NoGo sessions (30 positive and 30 negative stimuli in random order) the mean proportion of responses to positive and negative stimuli was 90.6±3.9 and 5.3±2.2, respectively (r=0.715±0.013). Before the next two sessions birds received intramuscularly 0.5 mg/kg of haloperidol or saline (control). Mean proportion of responses to positive and negative stimuli in haloperidol group was 63.33±12.72 and 0.00±0.00% respectively, while in control group it was 93.00±2.25 and 3.33±1.57%. The decrease in response to positive and negative stimuli was significant (P=0.020 and P=0.029 respectively). While all 10 control hens discriminated significantly the stimuli, 4 out of 10 hens after haloperidol treatment did not discriminate positive and negative stimulus. Our preliminary data show that the performance of laying hens in the Go/NoGo task decreased after the dopamine antagonist haloperidol. This work was supported by APVV-0047-10 and VEGA-2/0192/11.

Use of conditioned place preference test to determine how fish appraise positive and negative stimuli

Millot, Sandie, Cerqueira, Marco, Castanheira, Maria F. and Martins, Catarina, CCMAR, Centro de Ciências do Mar, Aquagroup, Universidade do Algarve, Campus de Gambelas, 8005-139 Faro, Portugal; smillot@ualg.pt

The large individual variation in stress responses does not only depend on the situation to which the individual is exposed, but also on the cognitive evaluation that the individual makes of the situation. The goal of this study was to understand how sea bream (*Sparus aurata*) appraised their environment by assessing the behavioural profiles of fish exposed to positive and negative stimuli. To this aim, individual fish was tested in a conditioned place preference tank divided in three compartments: one central compartment (start chamber) and two lateral compartments, one with white walls and the other with black spots. During the habituation phase (1 day), fish were placed in the start chamber for 10 min and afterwards allowed to swim freely throughout the whole tank during 20 min in order to determine its initial preferred side (time spent >14 min). During the training phase (3 days), fish were presented either with the negative (sudden introduction of a net in the water in the initial preferred side) or the positive stimulus (release of one pellet in the initial less preferred side) 20 times during 10 min. The test phase (1 day) consisted of the same procedure as the habituation phase. The behaviour of each individual was recorded by video camera and analysed with the Lolitrack 2.0 software. The preliminary results showed that the fish submitted to positive stimulus increased significantly the time spent (mean ± SD, +43±35%; $P<0.05$) and the distance moved (+12.7±15.7 m; $P<0.05$) in the stimulated zone. However, fish submitted to the negative stimulus did not show any changes in their swimming pattern. The protocol tested here can be used to investigate positive appraisal in fish but further improvements are needed to address negative appraisal in this fish species.

Do hens view photographs as substitutes for real objects?

Railton, Renee[1,2], Foster, Mary[2] and Temple, William[2], [1]Southern Research and Outreach Centre, University of Minnesota, 35838 120th St, Waseca 56093, USA, [2]University of Waikato, Psychology, Private Bag 3105, Hamilton 3240, New Zealand; renee.railton79@gmail.com

When studying animal welfare and perception, researchers have often used artificial two-dimensional stimuli (e.g. photos, videos and computer images) as substitutes for real animals or objects. However, there is little research showing whether animals interpret such stimuli as being the same as the three-dimensional objects they represent. This study investigated whether six hens could transfer a discrimination of real objects to photographs of those objects and vice versa. Three hens were trained, using a forced-choice discrimination procedure, to discriminate pairs of real objects that varied in either colour or shape (e.g. peck left key if shown circle, peck right key if shown triangle) until their accuracy reached at least 85% correct over 5 sessions. Three hens were trained to discriminate between photographs of the objects. All hens were then tested to see if discrimination transferred to the alternate stimuli they were not trained with (either photos or objects). Percentage correct remained high (above 85%) during test sessions with the differently coloured stimuli showing that the hens responded similarly to the objects and photographs (repeated measures ANOVA, ($F_{(2,10)}=2.528$, $P>0.05$). However, when the stimuli differed only in shape none of the hens showed transfer of discrimination to the alternative stimuli (ANOVA, ($F_{(3,15)}=5.795$, $P<0.05$) and percentage correct ranged between 48 and 85%. It was concluded that, when colour cues are removed, hens do not respond to objects depicted in photographs the same way they do to the real objects. Thus, caution is required if two-dimensional images are to be used as substitutes for real objects in welfare research.

Social and individual learning by laying hens in a feeding situation

Wichman, Anette, Swedish University of Agricultural Sciences, Department of Animal Environment and Health, P.O. Box 7068, 75007 Uppsala, Sweden; Anette.Wichman@slu.se

Hens have been shown to be socially influenced in their feeding behavior and this study investigated if they are more prone to follow an individual which has previously been successful in accessing food. Adult laying hens (LSL) were housed in eight groups with five to eight birds per floor pen which contained perches, nestboxes, litter and had ad lib access to food and water. Three birds from each group were used for the experiment. Training and testing was carried out in a separate arena (3.6×2.4 m) where six bowls with a lid were placed on the floor. During training the bowls contained a small amount of maize and the lids could be removed by an observer from outside of the arena. One of the birds in each trio had been randomly assigned to be the 'skilled' bird. When the skilled bird approached a bowl the lid was opened by the observer. Each trio was trained together seven times, on separate days. Birds were tested in pairs, once with each one of the other birds in the trio. During testing no lids were opened and no maize was provided. Observations of the frequency of pecks towards the bowls and the number of times a bird moved (minimum five steps) towards the other bird was carried out for three minutes. During testing the skilled birds performed more pecks towards the bowls (Mann-Whitney U Test; P=0.001), but there was no significant difference in how much unskilled birds followed a skilled or unskilled bird (Wilcoxon Signed Rank Test; P=0.25). Thus, birds that had been selected to be skilled approached bowls in order to get access to food but there was no evidence that the unskilled birds differentiated between their skilled and unskilled conspecifics in their following behaviour.

Using eggs for non-invasive stress-assessment: potentials, pitfalls and promising parameters

Rettenbacher, Sophie[1], Groothuis, Ton G.[2] and Henriksen, Rie[2], [1]University of Veterinary Medicine, Veterinärplatz 1, 1210 Vienna, Austria, [2]University of Groningen, Behavioural Biology, P.O. Box 11103, 9700 CC Groningen, the Netherlands; Sophie.Rettenbacher@vetmeduni.ac.at

Identifying potential stressors is a prerequisite for improving husbandry conditions of domestic animals. In birds, quantification of corticosterone in eggs has been suggested as a non-invasive method for stress assessment. However, corticosterone measurement in chicken eggs cannot be used as a welfare indicator, due to analytical issues: Egg yolk and albumin contain large amounts of progesterone and other gestagens which bind to the corticosterone antibody and seriously confound quantification of corticosterone via immunoassays. However, as synthesis of reproductive hormones is influenced by glucocorticoids, we investigated whether circulating corticosterone affects the production of gonadal steroid hormones and their deposition into the egg. Adult laying hens (n=20) of two different strains (ISA brown and white Leghorn) were implanted subcutaneously with corticosterone releasing pellets that elevated plasma corticosterone concentrations over a period of nine days. Control females received placebo implants. Eggs were collected over 19 days (n=407). Progesterone, testosterone and estrogens were quantified in plasma and yolk. Data were analysed in hierarchical models testing the effects of treatment, time, strain, as well as treatment × strain and treatment × time interactions. Corticosterone-implanted hens of both strains had lower plasma progesterone (P=0.049) and testosterone levels (P<0.001) and their yolks contained less progesterone (P=0.019; mean±sem concentration 3156±74 ng/g yolk) and testosterone (P=0.01; 3±0.1 ng/g), compared to controls (3430±78 ng/g and 3.6±0.1 ng/g, respectively). The treatment also reduced egg weight (P<0.001; 57.6±0.4 g vs. 61.3±0.3 g in controls) and yolk weight (P=0.009). Plasma estrogen concentrations decreased in white Leghorns only (P<0.001) whereas yolk estrogens were unaffected in both strains (P=0.7; 2.3±0.1 ng/g vs. 2.0±0 ng/g in controls). Our findings show that in domestic chickens, elevated plasma corticosterone concentrations suppress gonadal steroid hormone synthesis. Stress in laying hens could therefore be assessed non-invasively via quantification of reproductive steroids in yolk but also via parameters such as egg and yolk weight. The Dutch Animal Experiment Committee approved this experiment (project 5124A).

Effect of autonomic blockade on heart rate and heart rate variability of dwarf goats (*Capra hircus*)

Siebert, Katrin, Tuchscherer, Armin, Bellmann, Olaf and Langbein, Jan, Leibniz Institute for Farm Animal Biology, Behavioural Physiology, Wilhelm-Stahl-Allee 2, 18196 Dummerstorf, Germany; langbein@fbn-dummerstorf.de

Measurements of heart rate variability (HRV) have been increasingly used to assess changes in the vagal-sympathetic balance as related to animal welfare. To evaluate HRV measures, the autonomic control of the heart was analysed after sympathetic and parasympathetic blockade in dwarf goats. Two kinds of blockers were administered; propranolol (0.05 or 0.25 mg/kg i.v.) for beta-adrenergic blockade to inhibit sympathetic tone to the heart, and atropine (0.01 or 0.05 mg/kg i.v.) for selective parasympathetic nervous blockade. Twelve juvenile female goats were used in the study. Half of the animals were treated at a time with saline and atropine/propranolol respectively. Heart beat activity was recorded before and at intervals of 15, 30, 45 and 60 min after application by using Polar heart rate monitor S810i (©Polar Electro). Heart rate (HR), standard deviation of RR-intervals (SDNN, suggested to reflect sympathetic activity), differences between adjacent RR-intervals (RMSSD, mirroring vagal activity) and the RMSSD/SDNN ratio (vagal-sympathetic balance) were analysed. Statistical analyses were performed using SAS 9.2. A generalised linear model with fixed effects of treatment, interval, replicate and all two-way interactions was fitted by applying the GLIMMIX procedure. Statistical analysis was performed at a 5% level of significance. We found a temporary increase of HR after application of atropine or saline control, but a decrease of HR after application of propranolol for 60 min. Simultaneously the propranolol application reduced the SDNN for 60 min compared to the control, while no effect of atropine on this measure was observed. RMSSD was reduced after atropine application for 30 min compared to saline control. Finally, we found a prolonged increase of the RMSSD/SDNN ratio after propranolol application but a transient decrease after application of atropine compared to the saline control. Results prove that HR, SDNN and RMSSD are informative measures of sympathetic and parasympathetic activity in dwarf goats.

Factors influencing heart rate variability during lying in dairy cows

Gutmann, Anke K.[1], Ledochowski, Kira[1], Špinka, Marek[2] and Winckler, Christoph[1], [1]University of Natural Resources and Life Sciences, BOKU, Department of Sustainable Agricultural Systems, Division of Livestock Sciences, Gregor-Mendel-Strasse 33, 1180 Vienna, Austria, [2]Institute of Animal Science, Department of Ethology, Přátelství 815, 104 00 Praha Uhříněves, Czech Republic; kira.ledochowski@boku.ac.at

Heart rate variability (HRV) is a promising measure to get insight into animals' inner state. Usually, HRV during specific situations is compared to baseline values during resting. However, variability of HRV during resting has not yet been scientifically determined. We investigated HRV of Holstein Friesian dairy cows throughout the daylight period on two to three consecutive days. HRV was measured using Polar RS800CX®, and behaviour sampled continuously from video recordings. Data were processed in Kubios HRV® and the following parameters determined: heart rate (HR), root mean square of successive differences (RMSSD), standard deviation of interbeat-intervals (SDNN), power spectral density estimate for high frequency band powers (HF 0.25-0.58 Hz) in normalized units (nu), and low frequency (LF, 0.04-0.25 Hz)/HF power ratio. The data set comprised in total 358 10-minute-intervals during lying (113 from 6 primiparous cows (mean per cow±SE=19±2.8) and 245 from 10 multiparous cows (27±3.7)). The effect of parity, rumination, daytime (morning, early-afternoon, and post-milking), and the interactions parity×daytime, parity×rumination and daytime×rumination on all HRV parameters was analysed using general linear mixed models with cow ID as random factor. HR differed significantly depending on daytime×parity (LSmeans±SE morning/early-afternoon/post-milking for primiparous vs. multiparous cows: 85±2.4/87±2.5/90±2.4 vs. 80±1.9/79±2.0/85±1.9, P<0.05), and ruminating×parity (non-ruminating/ruminating for primiparous vs. multiparous cows: 88±2.4/87±2.4 vs. 80±1.9/82±1.9, P<0.01). For SDNN, RMSSD and HFnu values differed significantly depending on daytime×parity (morning/early-afternoon/post-milking for primiparous vs. multiparous cows: SDNN: 20±2.8/14±3.1/8±2.3 vs. 14±1.9/18±2.1/10±1.8, P<0.05; RMSSD 24±3.5/15±3.9/4±3.0 vs. 13±2.4/17±2.7/6±2.3, P<0,05; HFnu 28±3.5/23±3.8/10±3.2 vs. 19±2.6/19±2.8/12±2.6, P<0.05). LF/HF power ratio was significantly different depending on daytime (morning vs. early-afternoon/post-milking: 9±2.7 vs. 12±2.8/15±2.6, P<0.01). These patterns may be due to some diurnal rhythm, milk production or stress related to udder fill or milking procedure, indicating higher (pre-)milking stress in primiparous cows. Independent of interpretation, time of day, rumination activity and parity should be considered when using baseline HRV values.

Cortisol and sIgA measurements after acute stress in puppies and adult police dogs

Chaloupkova, Helena[1,2], Svobodova, Ivona[2], Koncel, Roman[2], Hradecka, Lenka[2] and Jebavy, Lukas[2], [1]Institute of Animal Science, Department of Ethology, Pratelstvi 815, Praha Uhrineves, 10400, Czech Republic, [2]Czech University of Life Science, Department of Husbandry and Ethology of Animals, Kamycka 129, Praha Suchdol, 16021, Czech Republic; chaloupkovah@af.czu.cz

Behavioural puppy tests are routinely used to assess later working ability of police dogs in the Czech Republic. It has been shown that behaviour is linked to different physiological responses to stressful situations. The aim of the study was to find out whether cortisol and sIgA can be useful indicators of acute stress in young puppies and adult dogs. Seventeen puppies were tested at the age of 7 weeks according to a standard method used in the Czech Republic police breeding facilities. The 5 min-long test consisted of 10 tasks (e.g. noise, retrieval, tug of war). Ten adult females were subjected to 2 min-long training of defence. Sufficient amount of saliva was collected before testing as control samples and 20 min after the start of the challenge on a cotton swab held for 1 min in the mouth of the dogs. The concentration of both measurements was analyzed using ELISA tests. The differences in cortisol and sIgA concentration were tested using Mixed Model (PROC MIXED) in SAS. In puppies, we found an increase of cortisol concentration after behavioural testing compared to control samples ($F_{1,8}$=8.59, P<0.05). The sIgA concentration was not affected by puppy testing ($F_{1,14}$=2.87, ns). In both measurements no significant effects of sex and weight of puppies were detected. In adults, we found an increase of cortisol ($F_{1,9}$=7.43, P<0.05) and a decreases of sIgA ($F_{1,9}$=5.78, P<0.05) after defence training compared to control samples. The study showed that cortisol can be used as an indicator of acute stress in young puppies but not sIgA. In adults, decreasing of sIgA can indicate a response to the acute stress, but further research should be carried out to clarify the physiological responses after different types of stress stimuli.

Validating the accuracy of activity monitor data from dairy cows housed in a pasture-based automatic milking system

Elischer, Melissa F., Karcher, Elizabeth L. and Siegford, Janice M., Michigan State University, Department of Animal Science, 1290 Anthony Hall, East Lansing, MI 48910, USA; elischer@msu.edu

Behavioral observations are important in detecting illness, injury, and reproductive status as well as performance of normal behaviors. However, conducting live observations in extensive systems, such as pasture-based dairies, can be difficult and time consuming. Activity monitors, such as those developed for use with automatic milking systems (AMS), have been developed to automatically and remotely collect individual behavioral data. In the Lely A3 Astronaut AMS, each cow wears a collar transponder (LELY) for identification, which can collect data on individual activity. The aim of this study was to examine whether cow activity levels as reported by LELY are accurate compared to live observations (OB) and previously validated pedometers (IceQube (IQ)). Fifteen lactating Holstein cows with pasture access were fitted with LELY and IQ. Continuous focal observations (0600-2000) generated data on inactive (lying (I)) and active behaviors. Activity recorded by OB and IQ included walking (W) and standing (S), while IQ steps (ST) measured cow movement (i.e. acceleration). Active behaviors were analyzed separately and in combination (A) to ascertain exactly what behavioral components contributed to calculation of LELY 'activity'. Pearson correlations for each cow were performed between LELY and: OBW (LOBW; 0.738±0.036), OBS (LOBS, 0.389±0.126), OBI (LOBI; -0.517±0.108), OBA (LOBA; 0.486±0.112), IQST (LIQST; 0.836±0.030), and IQA (LIQA; 0.542±0.097). Correlations were analyzed with general linear models. LOBW was similar to LIQST and LIQA (P>0.05), while differences were observed between LIQST and LIQA (P<0.05). LOBW, LIQST, and LIQA were significantly different (P<0.05) from LOBA and LOBS while no differences were observed between LOBA and LOBS (P>0.05). LOBI was significantly different (P<0.0001) from all correlations. These data suggest LELY fails to capture standing and lying behaviors, but does accurately reflect cow walking.

Aggressive behaviours of pigs analysed through video labelling techniques

Ismayilova, Gunel[1], Oczak, Maciej[2], Costa, Annamaria[1], Sonoda, Lilia T.[3], Viazzi, Stefano[4], Fontana, Ilaria[1], Fels, Michaela[3], Vranken, Erik[2,4], Hartung, Jörg[3], Berckmans, Daniel[4] and Guarino, Marcella[1], [1]The University of Milan, Department of Veterinary and Technological Sciences for Food Safety, via Celoria 10, 20133 Milan, Italy, [2]Fancom Research, Industrieterrein 34, 5981 NK Panningen, the Netherlands, [3]University of Veterinary Medicine Hannover, Foundation, Institute for Animal Hygiene, Animal Welfare and Farm Animal Behaviour, Bünteweg 17p, 30559 Hannover, Germany, [4]Katholieke Universiteit Leuven, M3-BIORES: Measure, Model & Manage Bioresponses, Kasteelpark Arenberg 30, 3001 Leuven, Belgium; gunel.ismayilova@unimi.it

A certain degree of aggressive behaviour is a normal part of pigs' behavioural repertoire to establish a dominance hierarchy. However, the current practice of mixing pigs and the rather barren environment in intensive housing conditions are recognised to increase the level of aggression between pigs. The objective of this study was to present the labelling techniques of video images as a method to quantify and describe pigs' aggressive behaviours. 'Labelling' could be defined as act of affixing labels, classification to animal's behaviour. The experiment was carried out at a commercial farm on a group of 11 male pigs weighing on average 23 kg, kept in a pen of 4×2.5 m. The video recordings were registered during 3 days after mixing and limited to 8 hours in total (day 1: 2 h, day 2: 3 h, day 3: 3 h). Labelling included observation of every single interaction on the video images frame by frame (11 frames per second) to determine the duration of each aggressive interaction and to describe the pig's behaviour and body positions in the early phase of aggression. Labelling took approximately 70 hours to be completed. As a result, 177 aggressive interactions were identified. The duration of most aggressive interactions (41.24%) was short, from 1 to 5 sec. A total of 7 positions at the start of aggressive interaction were identified (P1-P7). The body positions mostly observed were 'inverse parallel-P7' (39.5%), 'nose to nose, 90° -P2' (19.77%), 'nose to head-P4' (13.5%). The most frequent initial behaviours were head knocking (34.46%) and biting of different body areas (29.4%). Head knocking was mostly observed in relation to P7 and P2 positions and biting was mostly resulting from P7 position. The presented technique allowed a detailed and precise quantitative analysis of pigs' behaviour at the start of aggressive interactions, as well as extraction of information about their exact duration.

Why do pigs waste water? Description of drinking behaviour of growing pigs provided with water nipples

Andersen, Heidi M.-L. and Herskin, Mette S., Aarhus University, Dept. of Animal Science, Blichers Allé 20, 8830 Tjele, Denmark; HeidiMai.Andersen@agrsci.dk

Despite a water waste of 25-50% of the volume of water used, water nipples are common in commercial pig production. The water wasted is a problem from an environmental point of view and might indicate reduced animal welfare as well. Thus, the waste may be caused by animals never learning the appropriate use of the drinker. The aim of this project was to describe how growing pigs use water nipples. Fifty-two pigs (20.5±1.7 kg), housed in groups of 3 or 10 pigs/pen/nipple, were used. A bite nipple was situated in a box inside the pen, equipped with two video cameras allowing sidewards and downwards recordings. All pigs were marked on the back and fitted with an RFID-tag. Water use, water waste, and visits were automatically recorded for four consecutive days. Four drinking bouts/pig/day describing the head posture, bout duration, and way of drinking were recorded. The average flow rate was 0.82±0.08 l/min. The pigs were fed a libitum and straw was provided as enrichment. On average, the water use was 4.99 l/pig/day, of which 34.6% was wasted, but with a large variation between animals. Six different ways of drinking from the water nipple were observed. Each pig drank from the water nipple in 2 to 4 different manners. However, the majority of the pigs (n=38) had a preferred way of drinking from the water nipple, which was observed for more than 60% of the drinking time. The preferred way to drink varied among the pigs. No clear relations were found between the way of drinking and the volume of water used. The results suggest that there is a need to redesign nipple drinkers for growing pigs – not only for environmental concerns – but also to ensure that all pigs meet their water needs. Further research is needed.

Acute stress in post-weaned pigs increases the detection threshold for sucrose

Figueroa, Jaime, Solà-Oriol, David, Pérez, José F. and Manteca, Xavier, Universitat Autònoma de Barcelona, Departament de Ciència Animal i dels Aliments, 08193 Bellaterra, Barcelona., Spain; figuejaime@gmail.com

It has been reported in some mammal species that chronic stress reduces the ability to detect low concentrations of sucrose; this process is known as anhedonia. The aim of the present study was to evaluate if pigs are able to change their preferences for a hedonic sweet solution due to an acute stress situation. A total of 240 post-weaned piglets (42 d-old) kept in 24 pens (10 pigs/pen) were randomly allocated in two groups: SG (stressed group) and CG (control group). SG pigs were exposed to an acute stress during 2 consecutive days by mixing animals of contiguous pens during 20 minutes. The ability to detect low sucrose concentrations in both groups were measured by performing a 30 minute double choice test (2 pigs/pen) between sucrose (0.5 or 1%) and tap-water after 15 min. of the stress periods. Sucrose concentrations were counterbalanced across pens to act as the first or second concentration test (day 1 and 2). Generalized Linear Model procedure of SAS® was used to analyse the data. Total consumption was not different between groups. CG piglets showed higher intakes of 0.5% and 1% sucrose solutions (153 vs. 57 ml; $P<0.05$ and 206 vs. 69 ml; $P<0.05$ respectively) over water. On the other hand, SG pigs did not show any difference between 0.5% sucrose and water (126 vs. 132 ml). However, at higher sucrose concentration (1%) they clearly preferred sucrose (258 vs. 124 ml). These results show that acute stress in pigs may increase their detection threshold to detect low concentrations of sucrose probably because of their impaired ability to experience pleasure as a result of a hedonic stimulus. Our results suggest that reducing weaning stress may increase the preference of palatable diets.

Time budget of dairy cows in feeding systems based on total mixed ration fed in conjunction with reduced grazed pasture

Hetti Arachchige, Anoma D.[1,2]*, Jongman, Ellen*[1,3]*, Wales, William J.*[4]*, Fisher, Andrew D.*[1,2] *and Auldist, Martin J.*[4]*,* [1]*Animal Welfare Science Centre, University of Melbourne, Parkville, Victoria 3010, Australia,* [2]*Faculty of Veterinary Science, University of Melbourne, Parkville, Victoria 3010, Australia,* [3]*Department of Primary Industries, 600, Sneydes Road, Werribee, Victoria 3030, Australia,* [4]*Department of Primary Industries, Hazeldean Road, Ellinbank, Victoria 3821, Australia; anomah@student.unimelb.edu.au*

In pasture-based dairying, feeding supplements as a partial mixed ration (PMR) on a feed-pad may offer advantages over feeding concentrates during milking. However, feed-pad feeding may affect grazing cows' key behaviours, particularly lying time, due to increased time off pasture. Lying time is known to affect both milk production and well-being. The objective was to quantify the time budget of grazing cows fed PMR or concentrate feeding during milking (control). Holstein-Friesian cows were divided into 4 groups of 32 animals balanced for age (4.5±1.63 yrs), body condition score (4.6±0.27), live weight (561±55.2 kg) and milk yield (6,896±950.8 kg/cow/year). Two groups of control cows were supplemented with wheat grain during milking and pasture silage during grazing whereas 2 groups of PMR cows received supplements as a mixed ration on the feed pad for 90-120 mins after each milking. Both diets were isoenergetic and equivalent in amounts. Each group of 32 was further divided into 4 sub-groups of 8 cows, randomly assigned to one of 4 rates of supplement feeding (8, 10, 12 and 13.5 kg DM/cow/day). Following a 14-day adaptation period, behaviour was observed every 10 min for 24 hrs over 4 days. Data were analysed using a REML mixed effect model. There was no ration or rate effect ($P>0.05$) on daily lying time and number of lying bouts per day. Daily grazing time was lower in PMR cows than control cows (4.0 vs. 5.2 hrs, SED=0.16, $P<0.05$) and declined with increasing level of supplement (5.0, 4.9, 4.5, 4.1 hrs, SED=0.15, $P<0.05$). Ruminating time was not affected by ration ($P>0.05$) but increased with increasing level of supplement (8.0, 8.7, 8.9, 8.9 hrs, SED=0.21, $P<0.05$). In conclusion, PMR feeding slightly changed the time budget of grazing cows without compromising their lying time which is a major behavioural indicator of welfare.

How does solid feed provision and display of abnormal oral behaviours affect salivary cortisol in veal calves?

Bokkers, Eddie[1], Webb, Laura[1], Engel, Bas[2], Berends, Harma[3], Gerrits, Walter[3] and Van Reenen, Kees[4], [1]Animal Production Systems, Wageningen University, P.O. Box 338, 6700 AH Wageningen, the Netherlands, [2]Biometris, Wageningen University, P.O. Box 100, 6700 AC Wageningen, the Netherlands, [3]Animal Nutrition, Wageningen University, P.O. Box 338, 6700 AH Wageningen, the Netherlands, [4]Livestock Research, Wageningen University and Research Centre, P.O. Box 65, 8200 AB Lelystad, the Netherlands; eddie.bokkers@wur.nl

Veal calves display abnormal oral behaviours (AOB) due to frustration from limited opportunity to chew and ruminate. We investigated effects of solid feed provision and AOB display on saliva cortisol response, as a possible measure of chronic stress. This study piggy-backed on an existing nutrition study. Group-housed bull calves (n=48) were fed four feed amounts on top of milk replacer (0, 9, 18, 27 g DM/kg$^{0.75}$/d).The solid feed comprised of 50% concentrates, 25% maize silage and 25% wheat straw on dry matter basis. At 16 wk, calves were moved to individual metabolic cages and restrained for 10 days. Saliva was collected at -6 h (baseline), +40 min, +80 min, +120 min and +48 h relative to restraint in cage. At 24 wk, calves were moved to metabolic cages again and received catheters one day later. Saliva was sampled at -24 h (baseline), -1 h, +40 min, +80 min, and +120 min relative to catheter implantation. AOB were recorded every week (from 8 to 24 wk of age) in the home pens, using instantaneous scan sampling at 5 min interval for 30 min every 2 h from 06:00 to 19:00, once a week. Restraint at 16 wk led to lower cortisol peaks (P<0.01) and faster return to baseline levels compared to catheterisation at 24 wk. At +40 and +80 min, saliva cortisol was lower in calves fed 0 g solid feed (P<0.05). The lower cortisol response to stressors in calves fed no solid feed suggests increased chronic stress prior to moving. A negative relationship between cortisol baseline at 16 wk and AOB prior to 16 wk was found (P=0.028). This suggests displaying AOB could reduce baseline cortisol, which may reflect reduced chronic stress. No other relationships were found between cortisol and AOB. No differences were found between AOB 2 weeks prior and post restraint at 16 wk (P=0.105), thus restraint did not seem to affect stress-related behaviours in the home pens.

The effect of previous experience on the adaptation to headlocks of Holstein dairy cows

Krawczel, Peter D. and Hale, Julia M., The University of Tennessee, Animal Science, 2460 Morgan Circle Drive, 201E McCord Hall, 37996, Knoxville, TN, USA; pkrawcze@utk.edu

Moving dairy cows into new facilities requires an adaptation to novel environments. The objective of this study was to determine differences in behavior following the delivery of TMR between lactating cows with or without previous experience using headlocks. Forty-four Holstein dairy cows, from three different facilities, were assigned to 4 pens (n=11 per pen) within a new facility using headlocks as the only feed barrier. Two pens were comprised of naïve cows (originating from facilities using only post-and-rail feed barriers; parity=1.0±0.0; body weight=525.2±13.5 kg; DIM=234±8; milk production=25.8±0.2 kg/d) and two pens comprised of experienced cows (originating from a facility with only headlock feed barriers; parity=1.4±0.1; body weight=661.2±22.1 kg; DIM=328±11; milk production=18.6±0.2 kg/d). Their response was assessed from the mean percentage feeding, lying, standing in the feed alley, or other using 10-min scan samples collected via direct observation. Data were collected during 2 h after the morning and afternoon delivery of feed for 7 d. Data were log-transformed (backtransformed means reported) due to lack of normality and analyzed with the mixed procedure of SAS using repeated measures. More naïve cows engaged in feeding (47.3±1.0%) than experienced cows (29.4±1.0%; P<0.001). However, a greater percentage of experienced cows engaged were lying (48.3±1.0) and standing in the alley (15.4±1.0%) relative to the naïve cows (lying = 20.9±1.0% (P<0.001) and standing = 13.7±1.0% (P=0.02)). Experience level had no effect on other behaviour (P=0.71). The initial hypothesis, previous experience would benefit cows (indicated by a greater percentage of cows feeding and a lower percentage standing in the feed alley following the delivery of TMR), was rejected. Experienced cows and naïve cows behaved differently following introduction to headlocks, but there was no indication of a benefit from previous experience. This suggests other factors, such as days in milk or milk production, may be more important during adaption to novel environments.

Habitat use of *Bison bonasus* in a low mountain range enclosure

Schmitz, Philip and Witte, Klaudia, University of Siegen, Chemistry-Biology, Research Group Ecology & Ethology, Adolf-Reichwein-Str. 2, 57076 Siegen, Germany; schmitz@biologie.uni-siegen.de

In the E+E-Project 'Wisente im Rothaargebirge' (Germany, North Rhine-Westphalia), a herd (1 bull, 6 cows) of European Bison (*Bison bonasus*) is managed in an 88 ha area in a low mountain range habitat with beech, spruce and alder forest, grasslands and creeks. The animals are prepared for the reintroduction as a free-living herd in an intensively used working forest. We investigated habitat use, foraging behaviour and habitat preferences of the animals. We measured the home range size during 19 month using minimum convex polygone (MCP) and kernel with fixed bandwith (kernel h_{ref}), as well as the factors that shape the habitat use. As the animals do not use the whole area during summer and are able to attain their required food, we conclude that an 88 ha enclosure is suitable for this herd regarding the nutritional and ethological requirements during summer. As additional feeding is provided during winter, the animals limit their home range size to a core area of <1.1 ha. We also tested for correlations between habitat use and environmental factors. We found the strongest correlations between ground coverage by grasses and total grass biomass available. The animals avoided beech forests, prefer spruce forests and open areas as well as the respective shrub habitats. We found no correlations with the overall shrub-coverage, although the animals feed on several taxa. We additionally categorized the area in eight habitat types and calculated the modified Jacob's index for each season (vegetation-time versus non-vegetation-time). This revealed the animals avoid beech forests and prefer spruce forests, storm damaged areas and grasslands. All other habitat types are used at random. The feeding site was used frequently in both seasons. This leads to the conclusion that the nutritional requirements are crucial for the habitat preference of *B. bonasus* and most important for shaping their home ranges.

Behavioural changes prior to parturition in dairy cows

Jensen, Margit B., Aarhus University, Dept. of Animal Science, Blichers Allé 20, P.O. Box 50, 8830 Tjele, Denmark; margitbak.jensen@agrsci.dk

The aim was to investigate behavioural changes before calving to assess if behaviour may indicate imminent calving. Twenty-two multiparous Holstein-Frisian cows with easy and unassisted calving were housed in individual calving pens and fitted with leg attached accelerometers collecting data on lying behaviour continuously during 96 h before calving. Continuous records of behaviour via video, starting 12 h before calving, were obtained from 12 cows. Lying time was reduced (998^a, 987^a, 968^a and 894^b (±39) min/24 h, P<0.001) and number of lying bouts was increased (square-root, untransformed values in brackets: 4.1^b (17), 4.1^b (17), 4.2^b (17), and 4.9^a (24) (±0.13) bouts/24 h, P<0.001) during the day before calving. To analyse changes in lying behaviour during the final 24 h before calving, corrected for diurnal variation, from each 2-h value before calving was subtracted the mean value for that time of day during the preceding 3 days (delta-values). The delta number of lying bouts increased during the last three 2-h periods prior to calving, (-0.41^a, 0.03^a, 0.02^a, -0.05^a, 0.08^a, -0.02^a, 0.07^a, 0.13^a, 0.05^a, 0.48^b, 0.68^b, 1.31^c (±0.015) bouts/h, P<0.001). During the last four 2-h periods prior to calving the duration of contractions increased (natural log, untransformed values in brackets: -2.72^a (0.06), -1.62^a (0.20), -0.52^b (0.59), 0.05^b (1.06), 0.68^b (1.97), 2.62^c (13.7) (±0.55) min/h; P<0.001). During the final 2-h period before calving the number of turning head towards the abdomen increased (square-root, untransformed values in brackets 1.44^a (2.07), 1.52^a (2.31), 1.93^a (2.07), 1.72^a (3.72), 1.71^a (2.92), 4.05^b (16.4) (±0.26) no./h, P<0.001), while the duration of feeding (1.33^{ab}, 1.37^{ab}, 1.50^{ab}, 1.84^{ab}, 1.13^{bc}, 0.63^c (±0.26) min/h P<0.05) and drinking (0.46^{ab}, 0.52^{ab}, 0.63^a, 0.66^a, 0.18^{bc}, 0.04^c (±0.14) min/h P<0.01) decreased. The results show marked behavioural changes during the last hours before calving and suggest that behavioural changes may be useful indicators of imminent calving.

Predictability of a stressor enables dairy cows to handle more effectively a mildly challenging situation

Duvaux-Ponter, Christine[1], Rigalma, Karim[1,2], Barrier, Alice[1], Ponter, Andrew A.[3], Deschamps, François[4] and Roussel, Sabine[1,5], [1]AgroParisTech, 16 rue Claude Bernard, 75005 Paris, France, [2]LUBEM, Université de Brest, 29200 Brest, France, [3]ENVA, 7 avenue du Général de Gaulle, 94700 Maisons-Alfort, France, [4]Réseau de Transport d'Electricité, 34 rue Henri Regnault, 92400 Courbevoie, France, [5]IUEM, Université de Brest, 29200 Brest, France; christine.duvaux-ponter@agroparistech.fr

Dairy cows are often exposed to stray voltage (<10 V). We evaluated if being repeatedly submitted to this mild stressor modifies the ability to handle a mildly challenging situation. Seventy-four cows were submitted for 8 weeks to an unavoidable and therefore predictable (PRED; n=23), or random and unpredictable (UNPRED; 36 h/wk; n=25) low voltage exposure (1.8 V; 3.6 mA) applied to their water trough, or to no exposure (control; n=26). Two drinking stalls per treatment were accessible at all times through electronic individual recognition. In week 6, the motivation to drink after 5 h of water-access restriction was assessed by releasing three cows at a time (one from each treatment) from a pen 17 m from their drinking stalls. The test was repeated the following day with a novel object (coloured plastic bottle) hanging in the middle of each stall entrance. On day 1, while less than 10% of the cows reacted to the voltage by a sudden movement of the head but continued drinking, PRED and UNPRED cows had a higher mean heart rate than controls (P=0.01 and P=0.02, respectively). No differences were observed in weeks 2 and 8. In week 8, UNPRED cows had higher milk cortisol concentrations than PRED cows (P=0.02) and tended to have higher milk cortisol concentrations than controls (P=0.07). In week 6, PRED and UNPRED cows showed the same latency to drink and water intake than controls but PRED cows were faster to interact with the novel object than control (P=0.01) and UNPRED cows (P=0.04). UNPRED cows interacted less with the novel object than controls (P=0.02) and tended to interact less with it than PRED cows (P=0.08). Voltage exposure induced a transient acute stress response and, when predictable, appeared to reduce emotional reactivity and increase exploratory behaviour during a mildly challenging situation while, when unpredictable, voltage exposure decreased this exploratory behaviour.

Feeding time: signaled or not?

Nogueira, Selene, Abreu, Shauana, Macedo, Jaqueline, Costa, Thaise, Borges, Rogério and Nogueira-Filho, Sérgio L., Universidade Estadual de Santa Cruz, Applied Ethology Laboratory – DCB, Rod. Ilhéus-Itabuna km 16 Salobrinho, 45662-900, Brazil; seleneuesc@gmail.br

Some authors have suggested that an unpredictable environment, accompanied by some sort of signal for behavioral conditioning, could improve animals' welfare, promoting confidence and positive behaviors. We evaluated the effects of unpredictability with and without signaling by using feeding enrichment with 12 captive white-lipped peccaries (Tayassu pecari), categorized as a near-threatened species. The $A_1B_1A_2B_2A_3$ experimental treatment sequence ($A_{1,}$ A_2, A_3 – control – no change in spatial or temporal feeding routine; B_1 – non-signaled unpredictability – random feeding time and random feed location; B_2 – signaled unpredictability –, adding a whistle signal exactly when feed is provided) was followed. Each treatment lasted 10 days and each focal animal's activities were recorded for 5 minutes during feeding time from 8AM to 9AM, totaling 50 min per animal for each treatment. Time spent on exploratory and agonistic behavioral patterns on multiple days was totaled to create one record per individual in each experimental phase during feeding periods. We employed ANOVAs with repeated measures followed by post hoc Duncan test, when appropriate, using $P<0.05$ significance level. The animals spent more time exploring during both enrichment phases than in control ones (A_1: 249.7±187.5; B_1: 623.7±263.4; A_2: 359.9±188.0;B_2: 550.9±195.8; A_3: 520.5±215.4 Ps<0.01). There was no difference between unpredictable signaled and non-signaled enrichment phases (P=0.29). However, there was a carryover effect of B_2 to A_3 (P=0.56). During the different phases animals showed no differences in time spent on agonistic relationships (P=0.17) nor did live weight change (P=0.31). Peccaries showed no preference for signaled or non–signaled unpredictability, but signaled enrichment may contribute to animal welfare, because its prolonged effect on the next control phase increased explorations that mimic animals' natural behavior in tropical forests, where they cover up to 10 km per day while foraging.

Influence of exercise setting on heart rate response and body activity in dogs (*Canis familiaris*)

Fukuzawa, Megumi and Okubo, Miki, Nihon University, College of Bioresource Sciences, Dept. of Animal Resource and Sciences, Kameino 1866, Fujisawa, Kanagawa, 252-0880, Japan; fukuzawa.megumi@nihon-u.ac.jp

Lack of exercise can cause both dogs and humans to become overweight. Walking is thought to raise the body's activity levels and is a useful way of obtaining sufficient exercise. Although dog owners generally walk with their dogs, is the exercise that a dog obtains raising its activity levels as much as in its human owner? The purpose of the present study was to explore the heart rate (HR) responses and body activity levels in subjects undergoing exercise of three different exercise intensities (2, 4, and 6 km/h) in outdoor (field) and indoor (treadmill) settings. Ten healthy dogs (3 male, 7 female) wore a Polar HR monitor (Polar Electro Oy Corp., Finland) and an active mass measuring instrument (Calorism, Tanita Corp., Tokyo) throughout testing in the field or on a treadmill for 1 km. There were 4 German shepherd dogs, 2 Labrador retrievers, 1 Golden retriever, 1 Border collie, and 2 mongrels. Maximum HR (bpm), cardiac cost (%), activity energy expenditure (kJ), and metabolic equivalent of task (MET) were measured in each test. An ANOVA was used to assess the effect of dog, exercise intensity, environment on each measurement item. A change in exercise environment from field to treadmill did not affect the maximum HR, cardiac cost, or activity energy in the dogs. However, with a change in exercise intensity from 2 to 6 km/h there were statistically significant changes in maximum HR and cardiac cost (Tukey, $P<0.05$). Cardiac cost was also positively correlated with MET (dogs: $r^2=0.09$). These results suggest that, the body's activity levels of dogs might be similar to humans.

Presence of abscesses as a welfare indicator in dairy goats: a preliminary study

Ferrante, Valentina[1], Battini, Monica[1], Caslini, Chiara[1], Grosso, Lilia[1], Mantova, Elisa[1], Noè, Lorenzo[2], Barbieri, Sara[1] and Mattiello, Silvana[1], [1]Università degli Studi di Milano, Dipartimento di Scienze Animali, via Celoria, 10, 20133 Milano, Italy, [2]AGER s.c. AGricoltura E Ricerca, Via Tucidide 56, 20134 Milano, Italy; Valentina.Ferrante@unimi.it

Aim of this research was to verify if the presence of external abscesses can be associated with behavioural changes and a general poor welfare condition in dairy goats, to support the possibility of using it as a welfare indicator for on-farm welfare assessment. In a group of 70 Saanen goats at the same lactation stage, reared on straw litter in a commercial intensive farm, we selected 8 goats with visible abscesses near the lymph glands under the skin ('A'=Abscesses) and 8 goats with no visible abscess ('H'=Healthy). All goats were fed *ad libitum* hay and concentrate during milking. Each group was observed directly (scan sampling every 2 minutes) for 3 days for 3 h/d (1 h in the morning, 1 h in the afternoon, 1 h in the evening). The main observed behaviours were: feeding, standing, lying, moving, self-grooming, ruminating. Behavioural data were expressed as percentage of scans. BCS (0-3) was recorded. Data were submitted to univariate ANOVA (SPSS v18). Most of the observed behaviours did not show significant differences between groups. Moving ('A'=4.2%±0.2, 'H'=2.8%± 0.3) and ruminating ('A'=36.0%±3.6, 'H'=25.0%±1.9) were significantly higher in 'A' goats ($P<0.05$). Feeding was higher in 'H' goats ('H'=50.2%±0.5, 'A'=35.0%±3.5; $P<0.01$). In agreement with that, BCS was slightly higher in 'H' goats ('H'=1.53±0.12, 'A'=1.19±0.18; $P=0.14$); 25% of 'A' goats showed a very poor body condition (BCS=0.5), possibly as a consequence of the lower time spent feeding. The reduced access to the feed trough possibly induced 'A' goats to eat straw litter, and this may be the reason for the higher percentage of time dedicated to rumination. The results seem to confirm that the presence of abscesses can be considered a valid 'animal based' indicator for on-farm welfare assessment. The authors thank the EU VII Framework program (FP7-KBBE-2010-4) for financing the Animal Welfare Indicators (AWIN) project.

Social perception of animal welfare in the Veterinary Medicine undergraduate program at a Mexican university

Mora, Patricia, Hernández, Ismael, Soto, Rosalba and Terrazas, Angelica, Facultad de Estudios Superiorer Cuautitlan U.N.A.M., Departamento de Ciencias Pecuarias y Departamento de Ciencias Sociales, km 2.5 Cuautitlan-Teoloyucan, San Sebastian Xhala, Cuautitlán Izcalli, Edo. México, 54714, Mexico; mormed2001@yahoo.com.mx

Varied strategies have been carried out to promote animal welfare around the world. However, success in every country is highly influenced by particular features of the human population in contact with animals. Some of these features are education, culture and background. The objective of the present study was to evaluate preliminary the knowledge on animal welfare in a community of a Veterinary Medicine undergraduate program, in order to implement appropriate training activities in the future. A questionnaire of 24 self-evaluative questions with dichotomous answer (yes or not) was used. 99 people were interviewed: 64 academics, 21 students and 14 assistants. The 24 questions were classified into 4 categories: general animal welfare knowledge; animal welfare legislation knowledge; awareness of other people's knowledge about animal welfare; and willingness to learn more about the topic. Obtained data were analyzed by proportions among interviewed groups within categories with Pearson chi-squared test. We obtained that most respondents were knowledgeable about animal welfare (students 91%, academics 97% and assistants 85%) and willing to learn more (students 95%, academics 100% and assistants 97%). Proportions were lower for concern about other people's knowledge (students 67%, academics 70%, and assistants 77%) and knowledge about animal welfare law (students 43%, academics 53%, and assistants 38%). Pearson chi-squared test revealed no significant differences between respondents in every category (P>0.05). We concluded that although respondents showed high knowledgeable in the animal welfare topic, it is important to take in consideration their low concern about other people's knowledge and about animal welfare law, in order to implement an appropriate training activities. Research supported by grant UNAM-DGAPA-PAPIIT: IN217012

Animal welfare education: developing 'learning objects' from research in animal welfare science

Langford, Fritha[1], De Paula Vieira, Andreia[2], Vas, Judit[3], Varella Gomes, Pericles[2], Molento, Carla F.M.[4] and Zanella, Adroaldo J.[1], [1]SAC, Animal and Veterinary Sciences, West Mains Road, Edinburgh, EH9 3JG, United Kingdom, [2]Universidade Positivo, 5300 Campo Comprido, Curitiba, Brazil, [3]Norwegian University of Life Sciences, Animal and Aquacultural Sciences, P.O. Box 5003, 1432 Ås, Norway, [4]Universidade Federal do Paraná, Animal Sciences, Rua XV de Novembro, 1299, CEP 80.060-000, Curitiba, Brazil; fritha.langford@sac.ac.uk

The Animal Welfare Indicators (AWIN) project has created the 'Global Hub for Animal Welfare'. The primary objectives of the 'Global Hub' are to provide easy-to-access information on existing animal welfare courses worldwide and to develop and host series of learning materials. The learning materials will be developed from a variety of animal welfare research both from within the AWIN project and best current knowledge of animal welfare. The Global Hub is aimed at veterinary education, training for the livestock production industries and community outreach activities around the world. These learning materials have been developed to be used either 'stand alone' or in modular form. As such, the educational model chosen for the learning materials to be hosted on the 'Global Hub' is the 'learning object' (LO). LO's are defined as 'online learning materials that are small, digital and can be broadly described as context independent, reusable and adaptable.' This highlights three educational dimensions required: 'Content', 'Technology' and 'Pedagogy'. The need for LO's to be small adds extra complication. How do we decide what is small? Information is available faster and in greater volume than before the internet and capturing the attention of learners is more difficult due to competition from other content providers. However, size will affect the level of complexity found within, shorter LO's are less complex than longer ones. If an LO is too small it will contain inadequate amounts of information to allow the learner to understand what they are learning, leading to an unrewarding learning experience. We will present examples of LO's from the AWIN project (1. the behavioural needs of the farrowing sow; 2. horse facial expressions) to help further the discussion on how optimal size and how best to provide information for animal welfare education.

Behavioural patterns and performance indicators in entire male pigs housed in visual or non visual contact with females

Fàbrega-Romans, Emma[1], Soler, Joaquim[1], Montasell, Joan[1], Brillouët, Armelle[2], Company, Núria [3], Puigvert, Xavier[3] and Dalmau, Antoni[1], [1]IRTA, Animal Welfare and Animal Genetics, Veïnat de Sies, s/n, 17121 Monells, Spain, [2]Chambre d'Agriculture d'Ille-et-Vilaine, Pôle Porc-Aviculture, Rue Maurice Le Lannou – CS 14226, 35042 Rennes Cedex, France, [3]Universitat de Girona, Escola Politècnica Superior, c/Mª Aurèlia Capmany, 61. Campus Montilivi, 17071 Girona, Spain; emma.fabrega@irta.cat

A voluntary agreement to stop piglet castration by 2018 has been reached in the EU. Entire male production appears as one alternative, but no boar taint and proper welfare should be guaranteed. The objective of this study was to evaluate behaviour and performance of entire male pigs housed either in contact or non with females. Forty females (F) and eighty entire males, crosses of (Landrace × Large White) × Duroc, were used. Pigs were allocated in groups of 10 in 12 single sex pens, with sight, sound and touch contact but not direct mixing with the same gender (Males-Males, MM) or opposite gender (Males-Females, MF). Behaviour was video recorded 4 times (O1-O4: 127, 134, 162 and 169 days of age). Scan sampling was used to assess activity, inactivity, exploration and feeding, whereas focal sampling was used to evaluate agonistic and positive interactions and sexual behaviour. Automatic feeders allowed to collect individual feed intake and pigs were weighed every three weeks to estimate average daily gain (ADG) and food conversion ratio (FCR). Data was analysed with Proc Genmod or Mixed with repeated measures of SAS. No significant differences were observed for MM and MF for the scan sampling observations. However, MF presented a higher number of aggressions in O1 (7.00 vs. 3.66, $P<0.001$) and a higher number of mounts in O2 (1.67 vs. 0.25, $P<0.001$) compared to MM. No significant differences between MM and MF in positive interactions were observed, although age affected this behaviour (18.3 vs. 2.6, O1 vs. O4, $P<0.05$). Females presented a lower ($P<0.05$) ADG (809.6 g/day) than MM and MF (985.9 and 959.2 g/day, respectively), and a higher ($P<0.001$) FCR (3.16, 2.59 and 2.55 for F, MM and MF, respectively) Therefore, the presence of females in visual contact with males increased some behavioural patterns in males, with potential consequences on welfare and puberty onset and no influence on performance.

Dose-response effects of lidocaine on piglet behaviour at castration

Courboulay, Valérie, Lanneshoa, Maï and Eustache, Ophélie, IFIP – Institut du Porc, Pig farming, BP 35104, 35 651 Le Rheu cedex, France; valerie.courboulay@ifip.asso.fr

According to our previous studies, the injection of a 2.0% lidocaine solution reduces pain at castration but is associated with deleterious effect such as convulsions for some piglets. The aim of the present study was to determine both the benefits and drawbacks related to the use of lidocaine solutions ranging from 0.5 to 2.0%. A 2.0% lidocaine solution was therefore diluted to obtain the three other solutions at 0.5, 1.0, or 1.5% concentration. Injection was performed ten minutes before castration (1 ml/testis). Within each litter, all males received the same solution and volume. Sixteen litters were used per treatment, corresponding to 81, 80, 79 and 75 piglets for concentrations 0.5, 1.0, 1.5 and 2.0%, respectively. Behaviour, vocalisations and the duration of the surgery were recorded during castration. Behaviour was also recorded every 2 minutes during the following hour using scan sampling. Increasing lidocaine concentration resulted in a significant reduction in anterior and posterior legs movements as well as escape attempts during surgery (72, 58, 53 and 29% of piglets tried to escape for concentrations 0.5, 1.0, 1.5 and 2.0%, respectively, $P<0.001$). The intensity of vocalisations decreased when concentration increased (109, 102, 99 and 93 dB, respectively, $P<0.001$). Duration of castration decreased continuously from 16.3 to 14.3 seconds but differences among treatments were not significant. After castration, differences among treatments were not significant regarding isolation, postures and pain related behaviour. Convulsions were observed for two and seven piglets with 1.5 and 2.0% lidocaine, respectively. A 1.0% lidocaine concentration should be recommended to reduce pain at castration, without risk of convulsion. At higher concentration, attention should be paid to bodyweight of the piglet.

Pain and pain alleviation in pigs – a producer perspective

Wilson, Rebecca L.[1]*, Holyoake, Patricia K.*[2]*, Cronin, Greg M.*[3] *and Doyle, Rebecca E.*[1]*,* [1]*Charles Sturt University, Boorooma Street, Locked Bag 58, Wagga Wagga, NSW, 2678, Australia,* [2]*Department of Primary Industries, Cnr Midland Hwy & Taylor St, Epsom, VIC, 3551, Australia,* [3]*University of Sydney, Shute Building (C-01), 425 Werombi Road, Camden, NSW, 2570, Australia; rewilson@csu.edu.au*

The ability of pig producers to identify and alleviate pain in the animals under their care is crucial to animal health and well-being. The aim of this study was to determine how pig producers identify and manage pain in breeding sows, weaners and grower/finisher pigs on their farms. Data were collected during face-to-face interviews with 16 pig producers in Victoria, Australia during August and September, 2011. Production types included farrow-to-finish (n=12), grow-out only (n=1), breeder only (n=2) and farrow to weaner (n=1) farms. The questionnaire used during the interview consisted of 36 open and closed questions focussing on pain and heat stress recognition and management by pig producers. Answers provided by interviewees were validated during the farm visit by the principal investigator. Producers described 17 behaviours expressed by pigs in pain. The most frequent was vocalisation (eight producers, 50%), followed by change in gait and lethargy/listlessness (four producers, 25%). Routine husbandry procedures (tail docking and ear notching) were perceived as 'slightly to very painful', but pain relief was not considered warranted in these situations. Anti-inflammatories/analgesic products were mostly used to treat foot and joint problems, (11 producers, 68.75%), followed by mastitis (six producers, 37.5%) and general injuries/inflammation/meningitis (five producers, 31.25%). There appeared to be some confusion between anti-inflammatories/analgesics and antibiotics by some producers, with three producers (18.8%) and two producers (12.8%) naming tylosin and amoxicillin antibiotics, respectively, as anti-inflammatories/analgesics. The results of this study suggest that not all pig producers implemented anti-inflammatories/analgesics to relieve the pain associated with common illnesses and injuries in the animals under their care. Educating producers on the more subtle behavioural indicators of pain and the correct application of anti-inflammatories/analgesics may assist them to identify pain earlier and alleviate it appropriately.

The influence of breed and head position during beak trimming on the welfare of trimmed hens and their flock mates

Van Rooijen, Jeroen, Retired, Churchillweg 37c, 6707 JB Wageningen, the Netherlands; jeroenvrooijenAkira@hotmail.com

Beak trimmers aim at a lower beak that, during the laying period, is equal in length to the upper beak, or slightly longer. They claim that in (brown) Isabrown Warren hens the trimmed lower beak grows faster than the trimmed upper beak while in (white) LSL hens the reverse is the case. Further they claim to be able to correct for this by a changed head position during trimming. The upper and lower beak of 1,080 Isabrown Warren hens and 1,080 LSL hens were, with a hot blade, trimmed at the base of the nostrils by a professional beak trimmer. Of each breed one half was trimmed in the Warren position and the other half in the LSL position.One half of these groups was trimmed at day 1, the other half at week 6. Of each breed 360 hens remained untrimmed. Hens were kept in compartments on the floor (45 hens per compartment). Results were compared with analysis of variance. Immediately after trimming and at 16 weeks and 42 weeks it was scored whether the lower beak is shorter. In Warren hens the lower beak indeed grew faster. However, in LSL hens the growth speed of upper and lower beak was equal. Beaks trimmed in the Warren position (shorter lower beak) showed at 16 weeks significantly more abnormalities. At 42 weeks plumage conditions of all treatments was better than in untrimmed hens but the differences between the different treatments were not significant. However, during the laying period more dead hens were found among LSL trimmed Warren hens. It is concluded that trimming in the LSL position improved the welfare of the trimmed hens (because less beak abnormalities) but that trimming Warren hens in the LSL position decreased the welfare of their flock mates (because there was more cannibalism).

Stress response and performance of beef male calves after different castration methods in extensive systems

Del Campo Gigena, Marcia, Soares De Lima, Juan M. and Hernández, Santiago, National Institute of Agricultural Research – INIA Uruguay, Animal Welfare and Meat Quality, Ruta 5 km 386, PC 45000 – Tacuarembó, Uruguay; mdelcampo@tb.inia.org.uy

The stress response and performance of young calves castrated by different procedures are not clear for extensive rearing conditions. Forty Hereford calves, aged 7 days, were allocated to four different groups: (1) handled but uncastrated (C, n=10); or castrated by (2) surgery with local anaesthesia (lidocaine, 8 ml; LID, n=10), (3) surgery without anaesthesia (S, n=10); or (4) rubber ring without anaesthesia (RR, n=10). Serum cortisol concentration was measured in blood samples taken at 0, 1.5, 6, 24 and 48 hours and then every seventh day for 2 months. Behaviour was directly observed by scan sampling, immediately after the procedure (Day 1) and for 2 consecutive days, and then every seven days for 2 months. Calves were weighed once a week and remained with their mothers grazing native pastures. Cortisol level increased in all groups at 1.5 hours, returning to basal values at 6 hours, except in LID. S and RR exhibited a higher frequency of pain-associated behaviours than LID during Day 1, without differences between them. During day 2, LID increased the frequency of abnormal lying and standing associated to head turning ($P<0.05$, proc glimmix, SAS), without differences with S. By contrast, RR decreased the frequency of pain related behaviours during day 2, suggesting a shorter acute response when surgery is not involved. In this experiment, prolonged pain after RR castration was not evident, and no differences were found in physiological and behavioural indicators between castrated and uncastrated calves, after 48 hours. Weight gain increased along the experiment without differences between groups ($P>0.05$, proc mixed, SAS). In conclusion, anaesthesia reduced pain behaviour after castration on Day 1; RR seemed to be a good alternative for young, unweaned calves reared under extensive conditions; and pain did not seem to last more than 48 hours with all the techniques evaluated.

The effects of post partum ketoprofen on feeding behaviour, body condition, shoulder sores and constipation in multiparous sows

Viitasaari, Elina, Hänninen, Laura, Heinonen, Mari, Raekallio, Marja, Peltoniemi, Olli and Valros, Anna, University of Helsinki, Faculty of Veterinary Medicine., Koetilantie 7, 00014 Helsingin yliopisto, Helsinki, Finland; elina.viitasaari@helsinki.fi

Farrowing is painful and may cause prolonged inactivity. This may lead to discomfort and poor appetite, impairing the maintenance of body condition, leading to shoulder ulcers and prolonged constipation. We performed a double blinded, randomized study on 40 sows. Sows were kept in farrowing crates during the nursing period. Half of the sows received ketoprofen (NSAID) for 3 days post partum (3 mg/kg im/day) and the other half saline (PLACEBO). Persons administering the treatments and doing observations were blinded until analysis. Feeding behaviour was assessed by the feed left in the through and scored as either normal (trough is empty) or feed refusal (feed left in the trough). Body condition score (BCS) was measured on day 0, day 14 and at weaning using a 5-point scale (1=thin, 5=fat). Constipation and occurrence of shoulder sores were followed daily (days 0-7). Constipation was scored on a 3-point scale (0=no faeces, 2=normal faeces) and shoulder sores on a 4-point scale (0=intact skin, 3=severe or large wound). The effect of treatment on the first day of feed refusal, number of constipation days and occurrences of shoulder sores were studied with ANOVA and BCS using repeated measures model. Results are given as mean±SEM. Feed refusal occurred later in NSAID than PLACEBO sows (8.3±0.6 vs. 3.6±0.6, $P<0.05$). NSAID sows maintained BCS better than PLACEBO sows (2.9±0.1 vs. 2.5±0.1, $P=0.05$). Day 14 BCS was higher for NSAID sows than PLACEBO ones (2.9±0.1 vs. 2.4±0.2, $P<0.05$), but did not differ at farrowing or weaning. Shoulder sores appeared sooner for PLACEBO than NSAID sows (6.0±0.5 vs. 4.4±0.5 days, $P<0.05$). NSAID sows showed less constipation days than PLACEBO sows (5.7±0.2 vs. 6.5±0.3, $P<0.05$). Ketoprofen treated sows maintained good appetite longer and their BCS tended to be higher until weaning than placebo treated sows. Ketoprofen treatment delayed shoulder sore appearance and reduced the number of constipation days.

Mitigation of pain following non-elective caesarean section in cattle using meloxicam

Barrier, Alice[1], Dwyer, Cathy[1], Haskell, Marie[1] and Goby, Laurent[2], [1]SAC, West Mains Road, Edinburgh, EH9 3JG, United Kingdom, [2]Boehringer Ingelheim Animal Health, 55218 Ingelheim am Rhein, Germany; laurent.goby@boehringer-ingelheim.com

Cows that have undergone a caesarean section are likely to experience post-surgical pain in the days postpartum. However, there has been little consideration given to managing such pain and no drugs are currently licensed for that purpose. The objective of the study was to investigate if administration of meloxicam (a long-acting NSAID) prior to surgery would mitigate postpartum pain and discomfort in beef cows, by investigating activity-related behavioural changes. 110 beef cows (55 primiparous, 55 multiparous) that underwent non-elective standardised caesarean section were recruited from 8 French veterinary practices (investigators). Surgery took place under local anaesthesia, achieved with a line block of lidocaine hydrochloride. At the start of surgery, cows received either meloxicam (Metacam® 20 mg/ml, 0.5 mg/kg bodyweight SC) (n=63) or a placebo (n=47) according to a blind randomised schedule. Pedometers (IceTag®, IceRobotics Ltd, South Queensferry, UK) were attached to each cow's hindleg after surgery and the cow's activity was monitored from 0 h (end of surgery) to 24 h *post partum*. Percent of time spent lying, number of steps and Motion Index™ (absolute measure of acceleration of the leg, representative of the general movement) were calculated for the following periods: 0-8h, 8-16h, 16-24h. GLM analyses were performed using the interaction between time and treatment as a fixed effect and cow nested within treatment and investigator as a random effect. Cows receiving meloxicam spent significantly more time lying in the first 24 h following surgery than the cows receiving placebo (+62.6 min, $P<0.05$). However, there were no differences between treatments in the number of steps taken by the animals and in their Motion Index ($P>0.05$). Preemptive administration of meloxicam in cows undergoing surgery resulted in cows probably resting more after surgery. The relationship between resting and pain and discomfort should be further explored.

Behavioural indicators of pain in horses undergoing surgical castration

Dalla Costa, Emanuela[1], Rabolini, Aurora[1], Scelsa, Anna R.[1], Ravasio, Giuliano[2], Pecile, Alessandro[2], Lazzaretti, Sara[2], Canali, Elisabetta[1] and Minero, Michela[1], [1]Università degli Studi di Milano, Dipartimento di Scienze Animali, Sezione di Zootecnica Veterinaria, Via Celoria 10, 20133 Milano, Italy, [2]Università degli Studi di Milano, Dipartimento di Scienze Cliniche Veterinarie, Via Celoria 10, 20133 Milano, Italy; emanuela.dallacosta@unimi.it

Finding reliable pain indicators is paramount for equine welfare and little is available on post castration behaviour modifications in the horse. We aimed to investigate reliability and changes over time of behavioural pain indicators shown by horses undergoing routine castration. Eight stallions of different breeds, 2 to 4 years old, were castrated with closed technique in general anesthesia. The subjects were placed in an observation box for 2 days before and 3 days after intervention. Their behavior was video-recorded from a distance for 15 min at each of the following intervals: before surgery, 4, 8, 16, 24 and 40 hours after. Two blind observers, using Solomon software, analyzed duration and frequency of pain related behaviours. Observers significantly agreed (Kendall's coefficient of concordance for k related samples from a continuous field, $W<0.05$) assessing some behaviours (agitation, reluctance to move, kicking the abdomen, lethargy, lowered head carriage, flank watching, rolling, attention and curiosity) but not others (apathetic glance, abnormal walking). Compared to basal, 8 h after castration, horses showed significantly more (Wilcoxon test) agitation, reluctance to move, lowered head carriage and flank watching ($P<0.05$). A discriminant analysis was performed on concordant behaviours and 100% of the observations were correctly classified by the model as basal or 8 h. 16 hours after intervention the behaviour of horses was comparable to basal condition. We conclude that inter-observer reliability of assessors evaluating pain behaviour cannot be assumed and should always be evaluated. Altered behaviours occurred predominantly 8 h post castration, suggesting that this time is critical for pain evaluation and treatment. The authors wish to thank the EU VII Framework program (FP7-KBBE-2010-4) for financing the Animal Welfare Indicators (AWIN) project and for providing funds for Emanuela Dalla Costa to present this paper.

Back soreness is common among 'healthy' riding horses

Aksnes, Frida and Mejdell, Cecilie M., Norwegian Veterinary Institute, PB 750 Sentrum, 0454 Oslo, Norway; cecilie.mejdell@vetinst.no

Back problems are not uncommon in horses. We wanted to find the prevalence of back soreness in a 'normal' population of riding horse and investigate probable risk factors: equipment and rider. In total, 73 horses (29 riding school horses, 34 private riding horses and 9 police horses) at nine premises passed the inclusion criteria. The back was carefully palpated by a veterinarian student in her last year, using a standardized pressure. Reactions were categorized as: none; mild; medium; and severe. The fit of the saddle was categorized as well fit, too wide or too narrow. The horses were ridden in all three gaits and position of head and neck was noted. 43.8% of the horses showed back soreness, mostly in the caudal area. 46.5% had a well fitted saddle, 32.4% a narrow saddle, and 22.5% a wide saddle. Poor fit was significantly associated with back soreness ($P<0.01$), and a narrow saddle was worse. The use of pads sometimes turned a good fit into narrow. Horses with a high and stretched neck, and horses ridden with a high degree of neck flexion had significantly more often sore backs than horses ridden with head and neck in a more 'natural' position ($P<0.01$). Horses ridden regularly by three or more persons tended to show less pain reactions. The proportion of back soreness found in presumptively healthy horses in daily use was surprisingly high, as was the proportion of saddles with poor fit. In contrast to our prediction back soreness was not more common among riding school horses. Variation in load might be beneficial and outweigh the effect of beginners with poor balance. Back soreness in 'healthy' riding horses is a welfare problem which should be addressed in a cooperation of veterinarian, saddle experts and riding teachers.

Authors index

I

J

K

L

S

T

Printed in the United States
by Baker & Taylor Publisher Services